THE TIMES

Fiendish Su Doku

THE TIMES

Fiendish Su Doku

compiled by Wayne Gould

First published in 2006 by Times Books

HarperCollins Publishers
77–85 Fulham Palace Road
London
W6 8JB

www.collins.co.uk

Reprint 7

The Times is a registered trademark of Times Newspapers Ltd

ISBN-13 978-0-00-723253-6
ISBN-10 0-00-723253-5

A catalogue record for this book is available from the British Library.

Printed and bound in Great Britain by Clays Ltd, St Ives plc

Contents

Introduction vii

Puzzles
Fiendish 1–190
Super Fiendish 191–200

Solutions

Introduction

Since its launch in *The Times* in November 2004, Su Doku has become one of the most popular features of the paper and an international phenomenon. In a world where time is apparently a precious commodity, it is a testament to the addictive power of the puzzle that so many people can't wait to tackle it on a daily basis and with such intense concentration. *The Times* books, once they appeared in the bestseller lists, haven't budged since, showing their huge popularity with the book-buying public.

In this latest collection from Wayne Gould, you confront the ultimate mental workout. There are 190 new Fiendish Su Dokus with 10 Super Fiendish puzzles to test you to the absolute limit. These are the hardest puzzles he produces. Remember, his puzzles require no guesswork: logic will lead you to a single solution.

A valuable tip from Wayne Gould: 'If you are writing too many pencil marks, it means you are not understanding how the puzzle works. You may be relying too much on mechanical procedures, without appreciating the underlying logic. If, in time, you can shake yourself free of written pencil marks, you will see the Sudoku puzzle for what it is – a thing of beauty!'

Fiendish

		9		6				8
	7		2	3				1
	6				8			
		5	4			1	8	
	9	1			5	3		
			3				1	
9				8	6		5	
7				1		6		

Fiendish

2

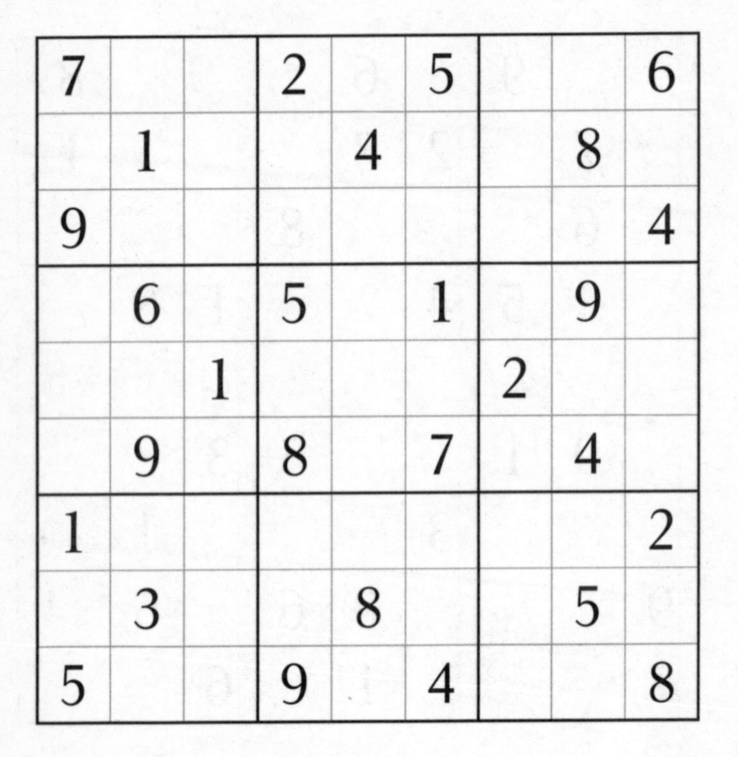

7			2		5			6
	1			4			8	
9								4
	6		5		1		9	
		1				2		
	9		8		7		4	
1								2
	3			8			5	
5			9		4			8

Fiendish

		2				9		
7								6
	6		4	8	3		2	
		4	9		6	5		
2								8
		1	3		8	4		
	2		1	9	4		3	
1								4
		3				1		

Fiendish

4

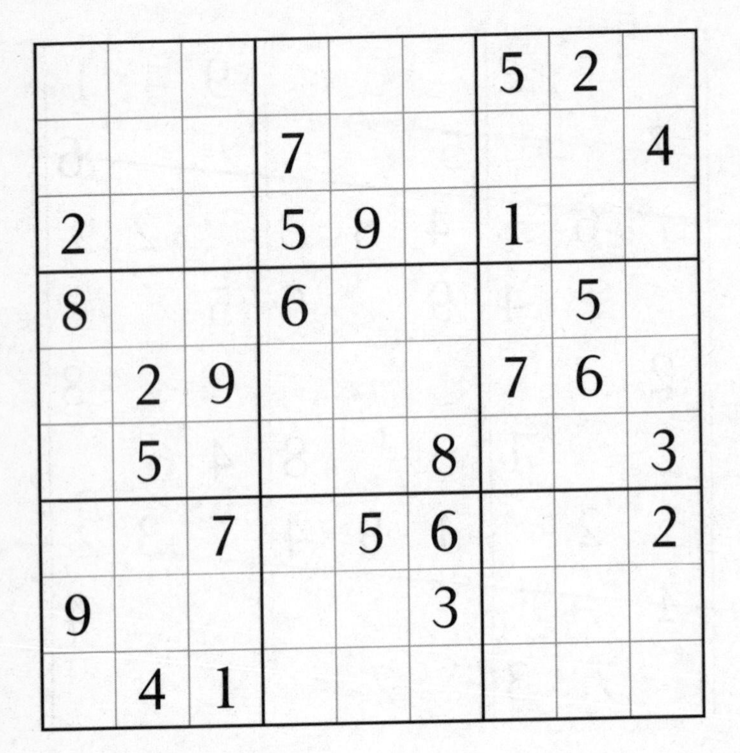

						5	2	
			7					4
2			5	9		1		
8			6				5	
	2	9				7	6	
	5				8			3
		7		5	6			2
9					3			
	4	1						

Fiendish

5

		2		6			4	1
			5			2		8
3					1			9
	9					5		
			9		4			
		7					6	
5			7					2
2		3			6			
1	7			2		8		

Fiendish

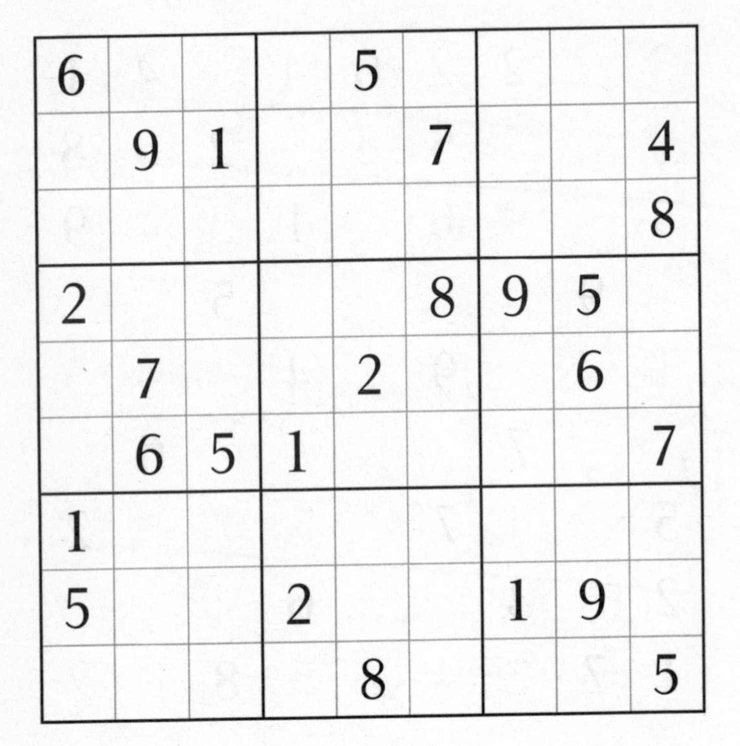

6				5				
	9	1			7			4
								8
2					8	9	5	
	7			2			6	
	6	5	1					7
1								
5			2			1	9	
				8				5

Fiendish

7

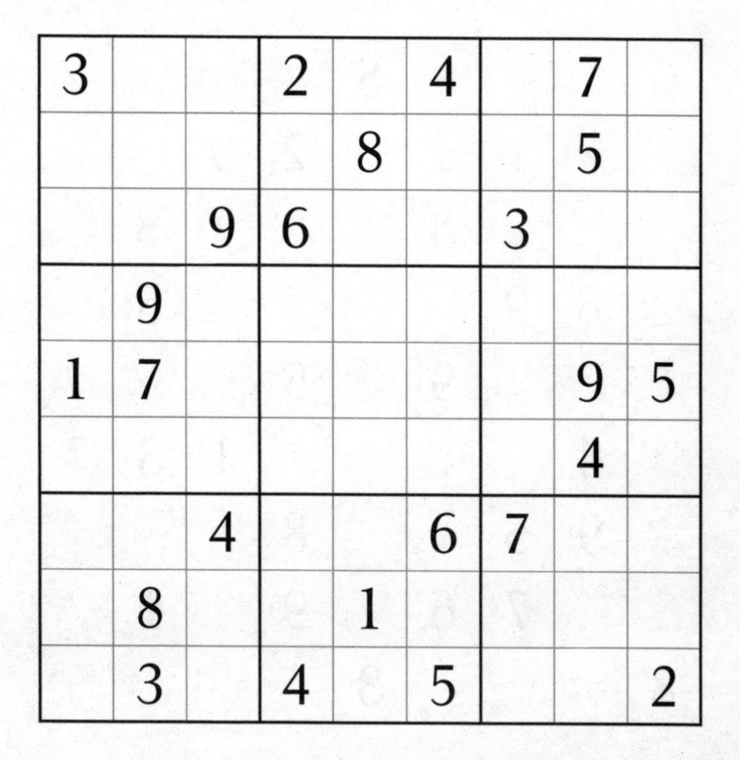

3			2		4		7	
				8			5	
		9	6			3		
	9							
1	7						9	5
							4	
		4			6	7		
	8			1				
	3		4		5			2

Fiendish

8

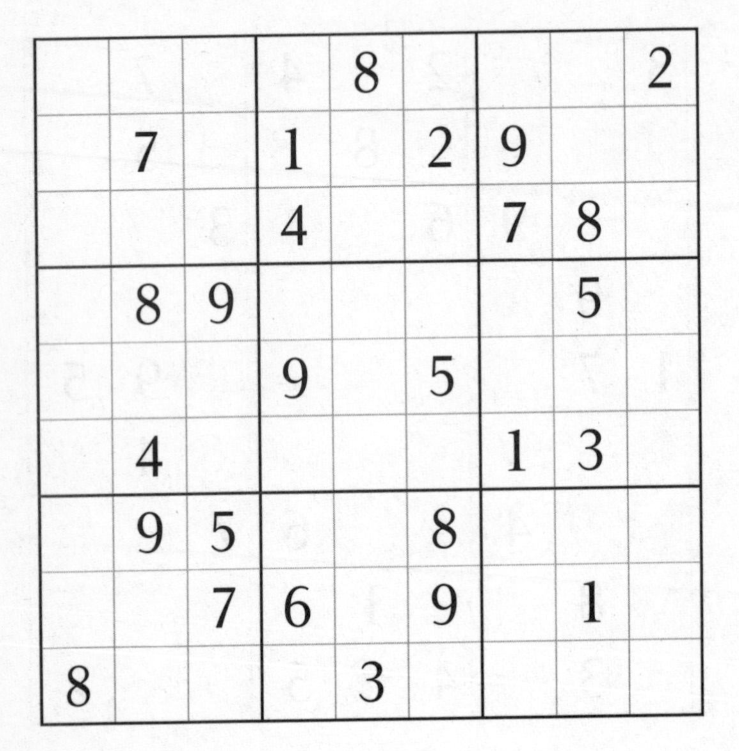

				8				2
	7		1		2	9		
			4			7	8	
	8	9					5	
			9		5			
	4					1	3	
	9	5			8			
		7	6		9		1	
8				3				

Fiendish

9

				1				
1					8	3		
			5		9	8	7	
		6				9	2	3
		5				1		
8	2	7				6		
	9	1	2		5			
		8	7					5
				8				

Fiendish

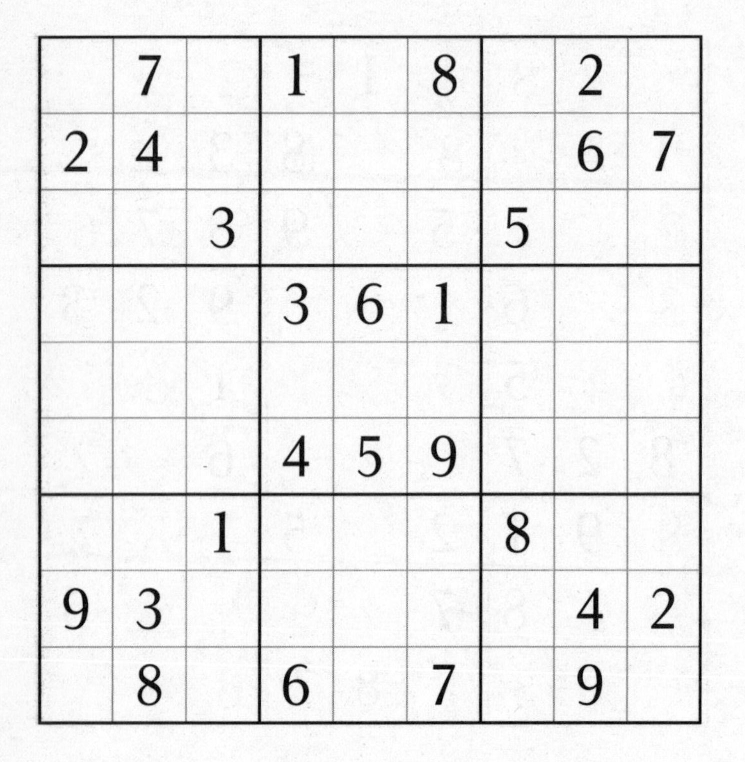

	7		1		8		2	
2	4						6	7
		3				5		
			3	6	1			
			4	5	9			
		1				8		
9	3						4	2
	8		6		7		9	

Fiendish

		8	4		5	2		
	7		8		6		9	
2								6
3								9
	4		3		1		8	
1								2
8								7
	2		6		9		1	
		3	1		7	6		

Fiendish

12

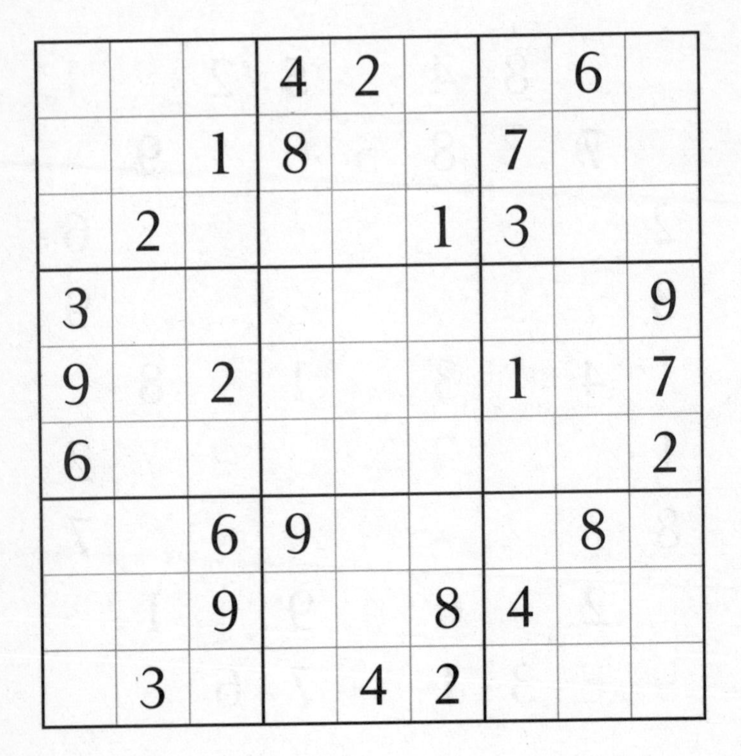

			4	2			6	
		1	8			7		
	2				1	3		
3								9
9		2				1		7
6								2
		6	9				8	
		9			8	4		
	3			4	2			

Fiendish

					3		2	1
	9	3		5			8	
				8		7		
3	7	9			5			
			3			2	7	5
		2		9				
	6			4		9	1	
9	8		1					

Fiendish

14

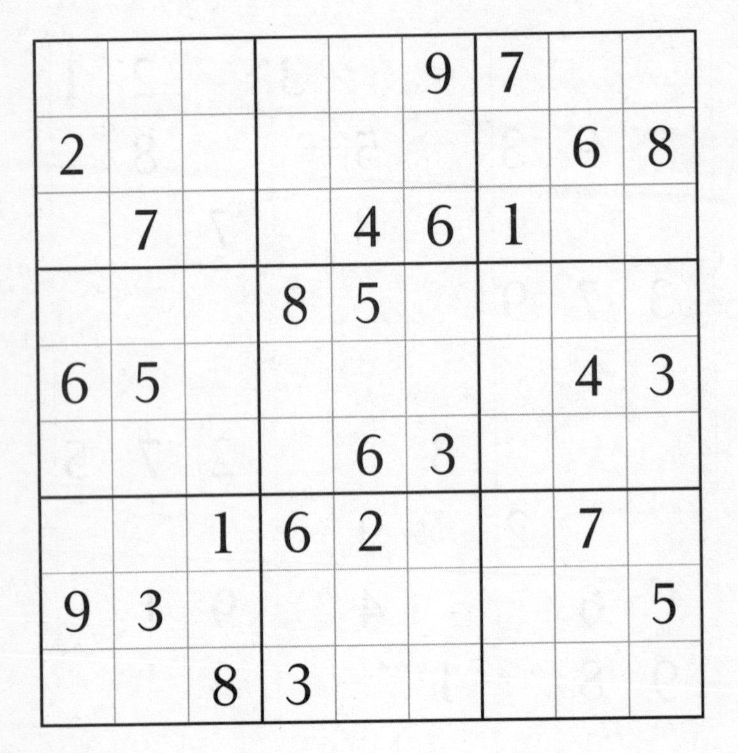

					9	7		
2							6	8
	7			4	6	1		
			8	5				
6	5						4	3
				6	3			
		1	6	2			7	
9	3							5
		8	3					

Fiendish

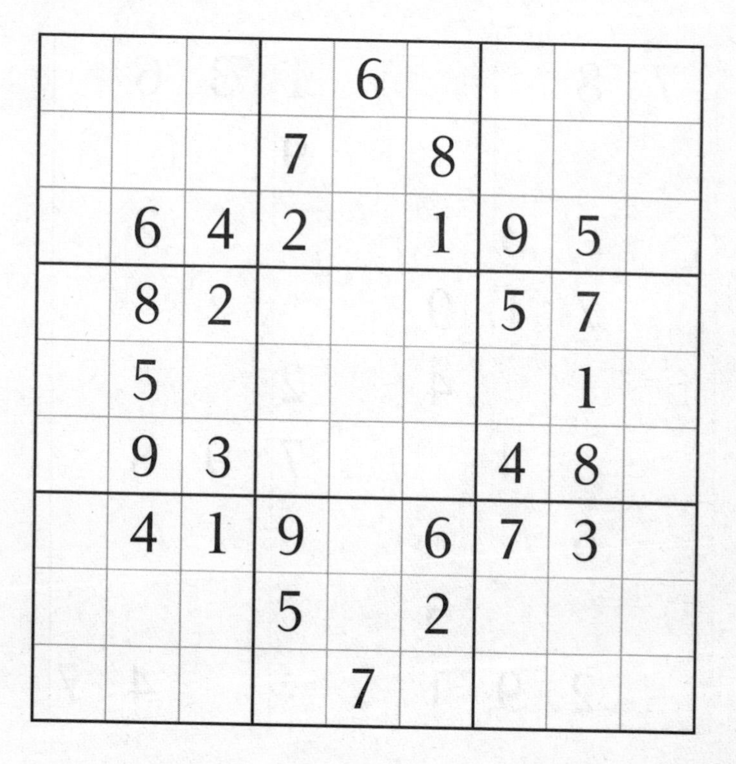

				6				
			7		8			
	6	4	2		1	9	5	
	8	2				5	7	
	5						1	
	9	3				4	8	
	4	1	9		6	7	3	
			5		2			
				7				

Fiendish

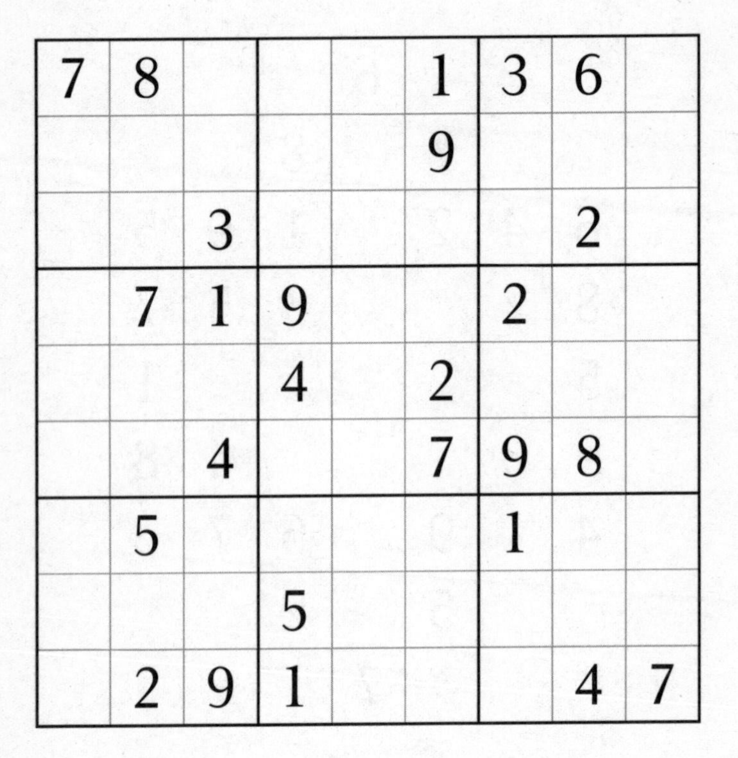

7	8				1	3	6	
					9			
		3					2	
	7	1	9			2		
			4		2			
		4			7	9	8	
	5					1		
			5					
	2	9	1				4	7

Fiendish

17

		6	4		5			
4	2		8			6		
	8	3						
	9				8			5
		4				2		
3			1				4	
						1	2	
		8			2		6	7
			3		7	8		

Fiendish

18

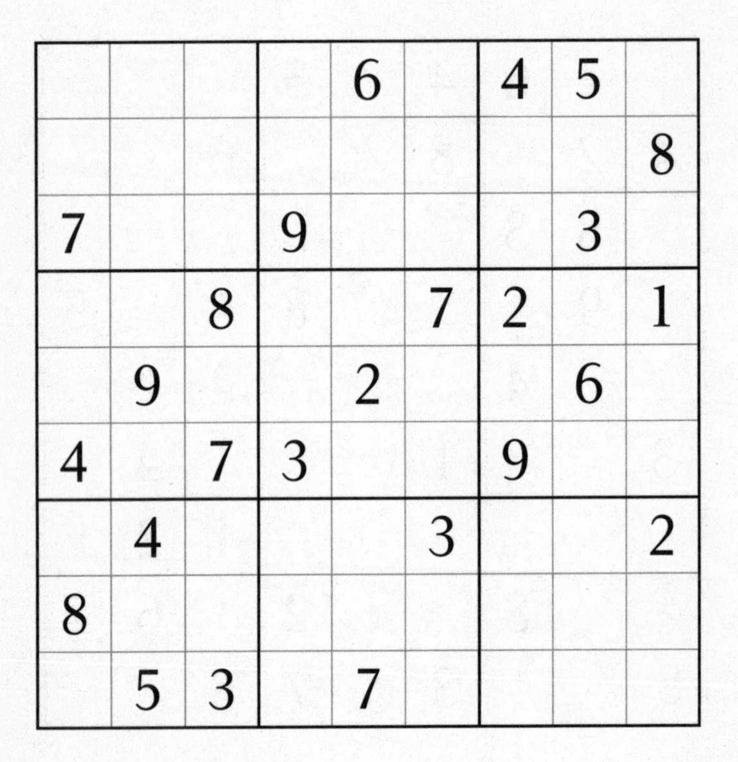

				6		4	5	
								8
7			9				3	
		8			7	2		1
	9			2			6	
4		7	3			9		
	4				3			2
8								
	5	3		7				

Fiendish

19

	2				4			6
		4		2		9		7
9	1							
		8	3					
		9		1		7		
					8	2		
							5	1
4		2		7		8		
8			5				7	

Fiendish

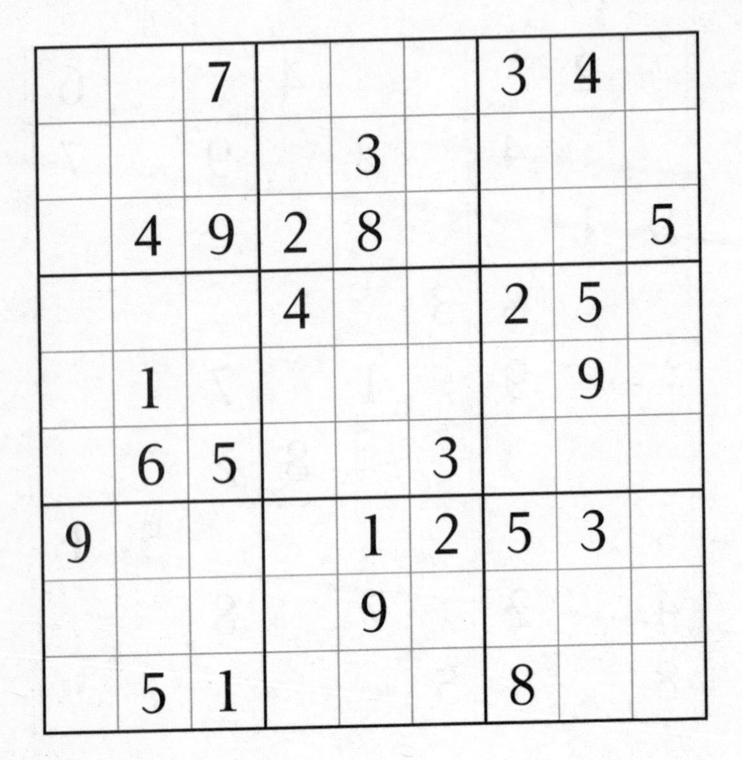

		7				3	4	
				3				
	4	9	2	8				5
			4			2	5	
	1						9	
	6	5			3			
9				1	2	5	3	
				9				
	5	1				8		

Fiendish

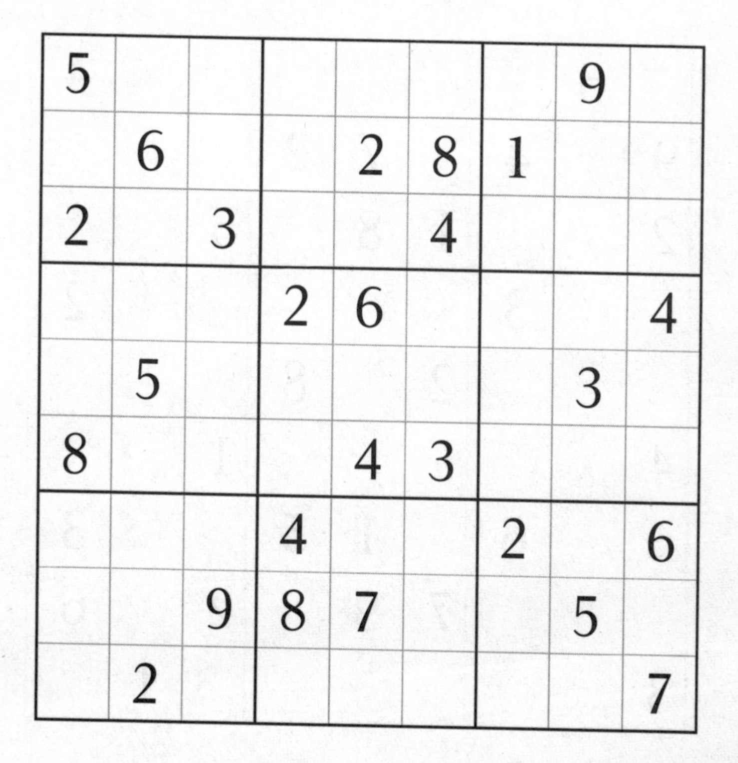

5							9	
	6			2	8	1		
2		3			4			
			2	6				4
	5						3	
8				4	3			
			4			2		6
		9	8	7			5	
	2							7

22

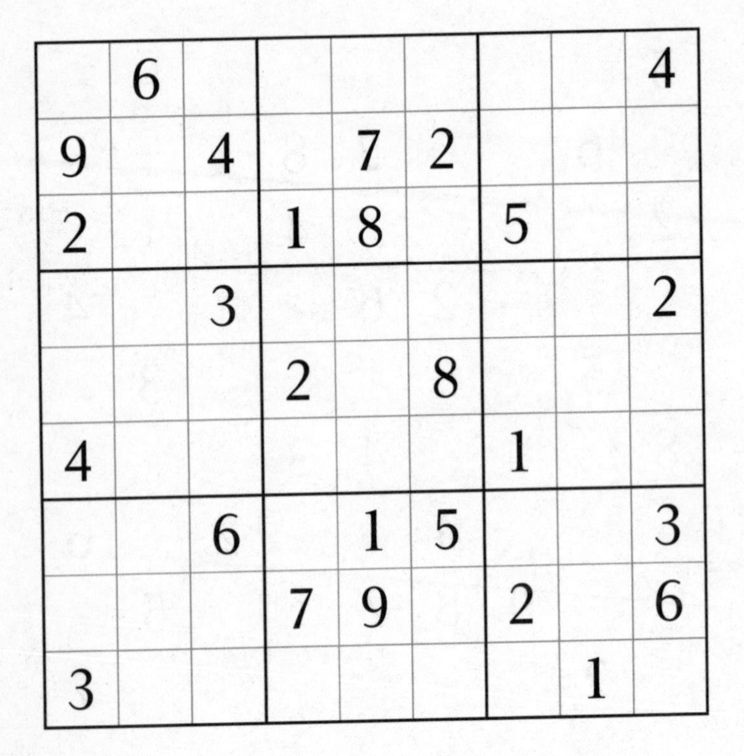

	6							4
9		4		7	2			
2			1	8		5		
		3						2
			2		8			
4						1		
		6		1	5			3
			7	9		2		6
3							1	

Fiendish

23

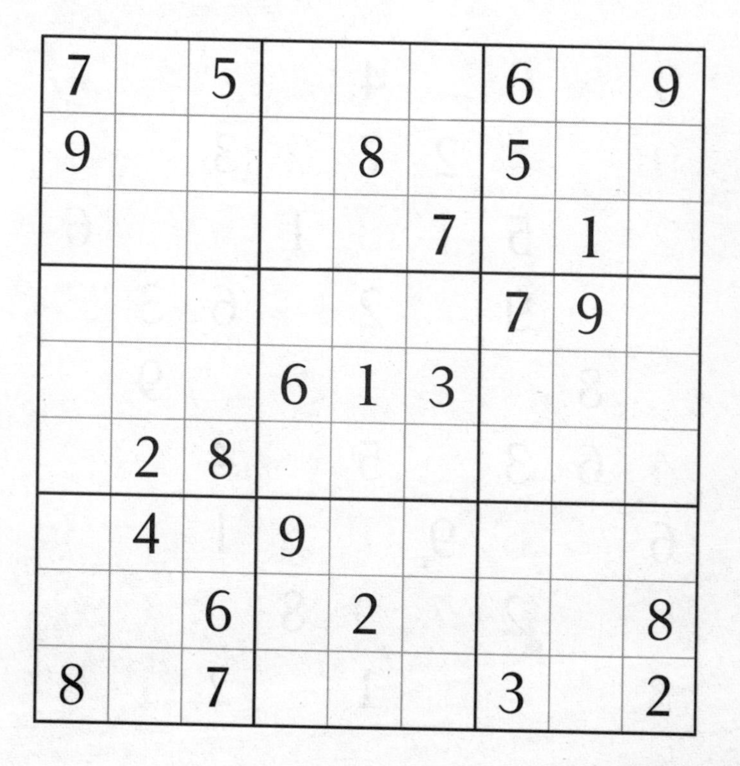

7		5				6		9
9				8		5		
					7		1	
						7	9	
			6	1	3			
	2	8						
	4		9					
		6		2				8
8		7				3		2

Fiendish

24

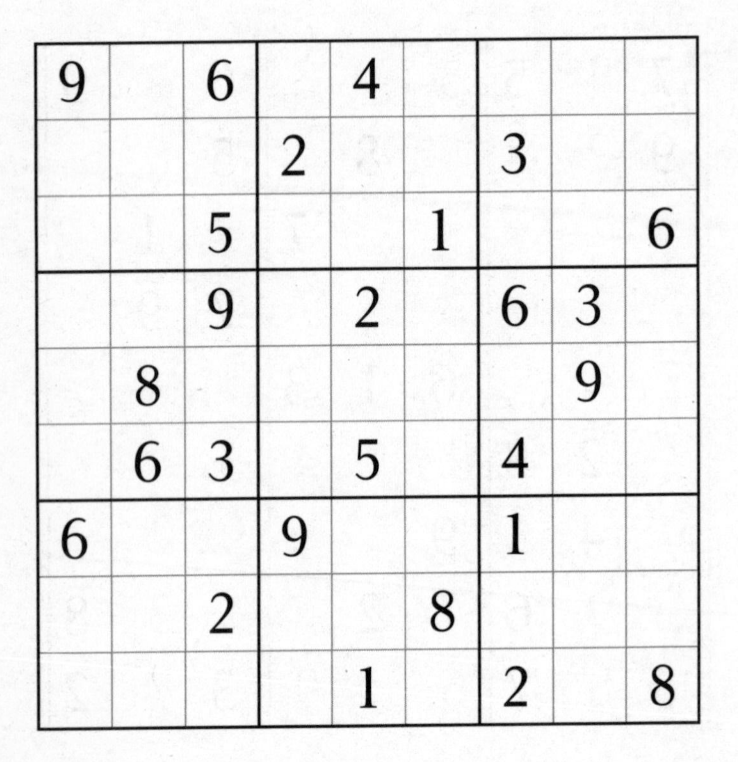

9		6		4				
			2			3		
		5			1			6
		9		2		6	3	
	8						9	
	6	3		5		4		
6			9			1		
		2			8			
				1		2		8

Fiendish

25

	4		5		3		2	
	2			4			8	
		3				6		
		6	3		7	2		
7								5
		8	1		4	3		
		2				9		
	9			2			3	
	8		6		9		7	

Fiendish

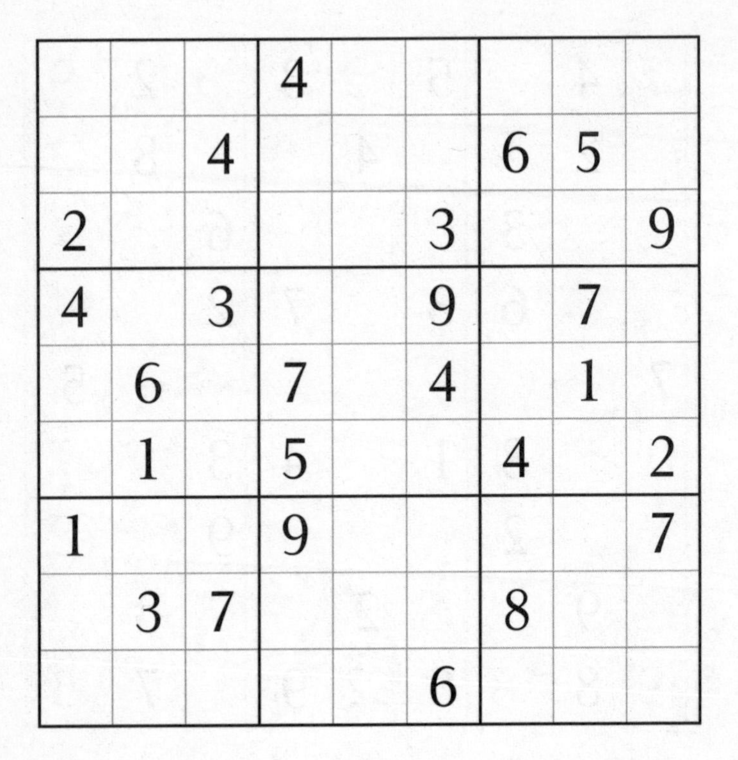

			4					
		4				6	5	
2					3			9
4		3			9		7	
	6		7		4		1	
	1		5			4		2
1			9					7
	3	7				8		
					6			

Fiendish

27

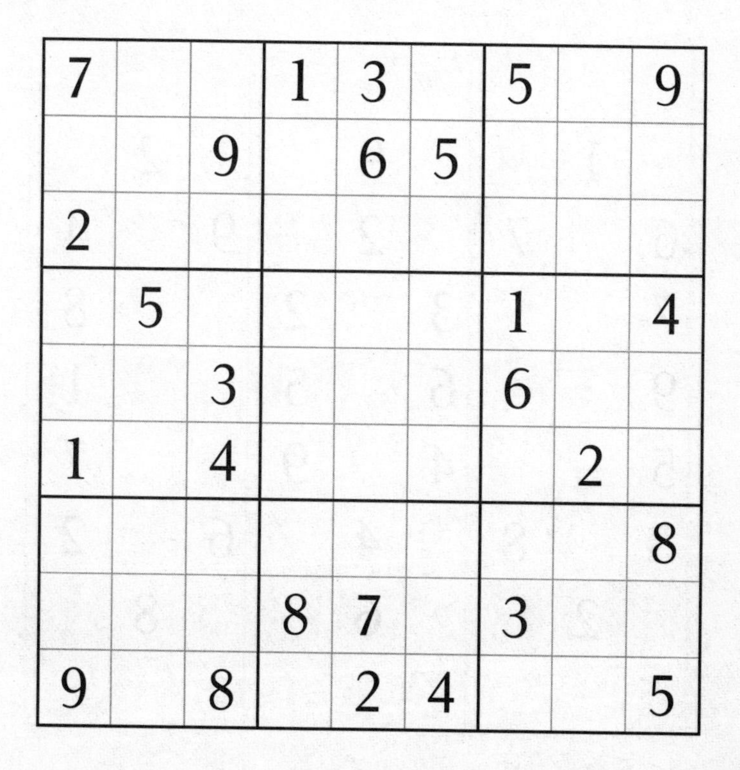

7			1	3		5		9
		9		6	5			
2								
	5					1		4
		3				6		
1		4					2	
								8
			8	7		3		
9		8		2	4			5

Fiendish

28

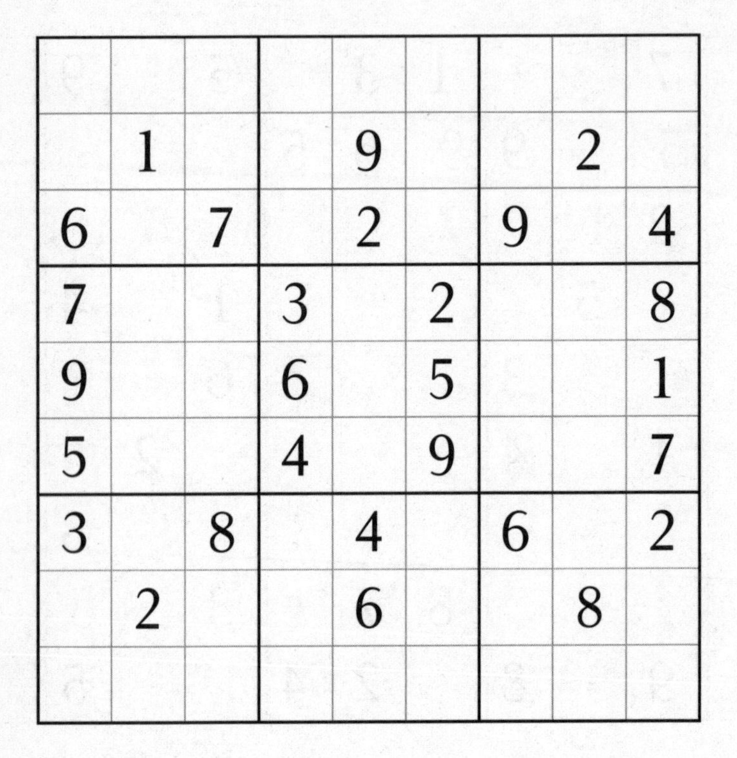

	1			9			2	
6		7		2		9		4
7			3		2			8
9			6		5			1
5			4		9			7
3		8		4		6		2
	2			6			8	

Fiendish

29

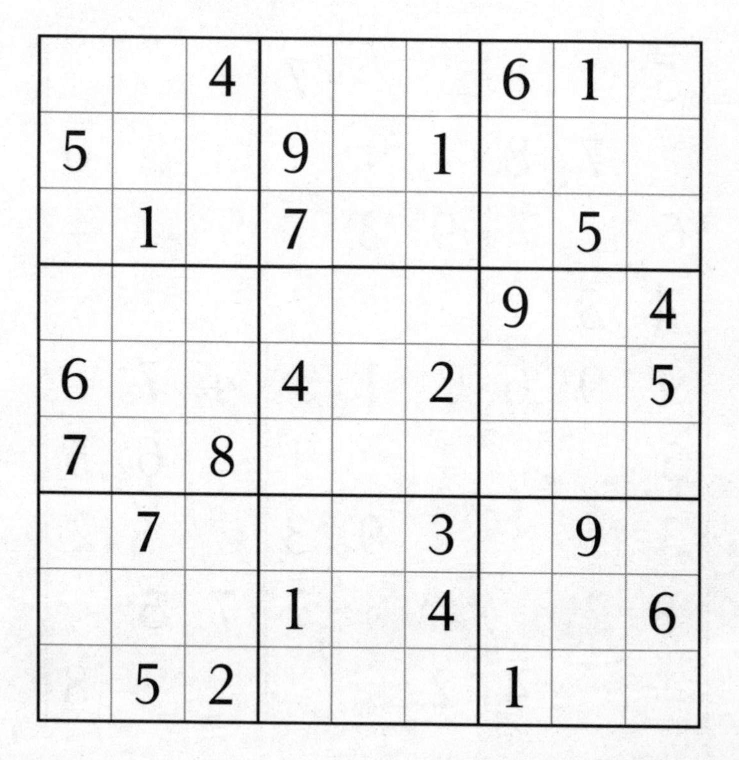

		4				6	1	
5			9		1			
	1		7				5	
						9		4
6			4		2			5
7		8						
	7				3		9	
			1		4			6
	5	2				1		

Fiendish

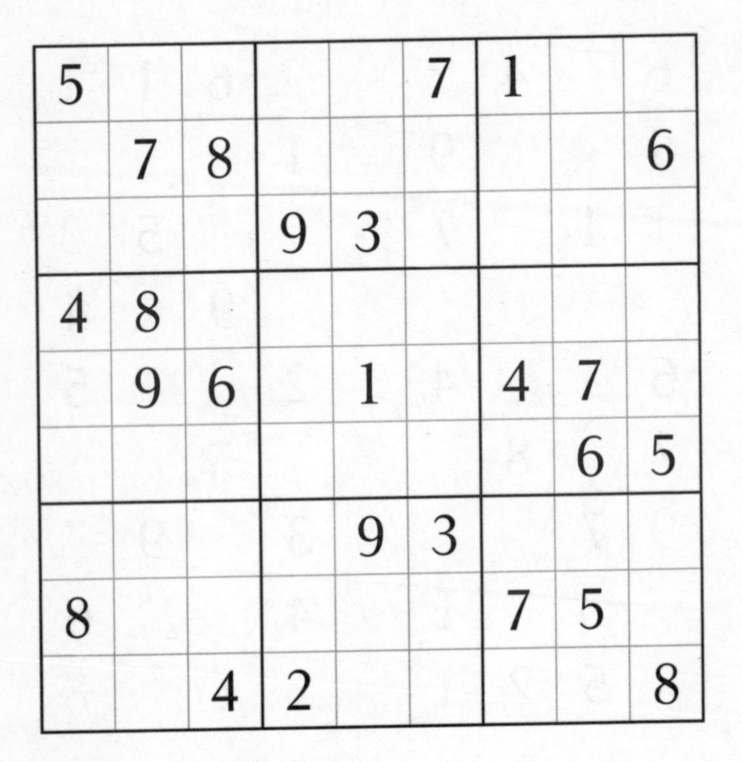

5					7	1		
	7	8						6
			9	3				
4	8							
	9	6		1		4	7	
							6	5
				9	3			
8						7	5	
		4	2					8

Fiendish

6			3		2			1
	4						5	
1				5				3
		6		3		8		
3	8						4	6
		4		7		5		
9				2				4
	3						2	
4			1		3			8

Fiendish

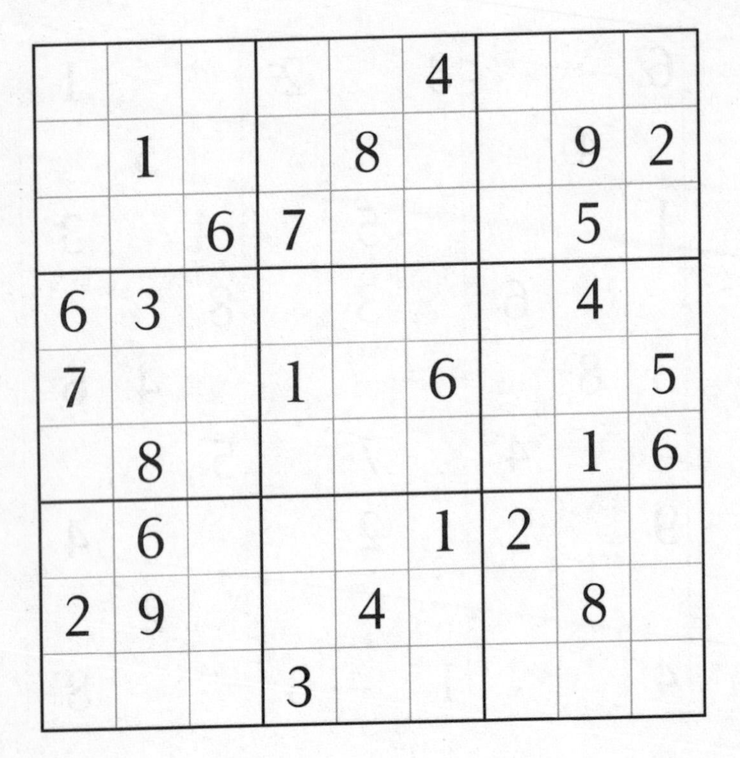

					4			
	1			8			9	2
		6	7				5	
6	3						4	
7			1		6			5
	8						1	6
	6				1	2		
2	9			4			8	
			3					

Fiendish

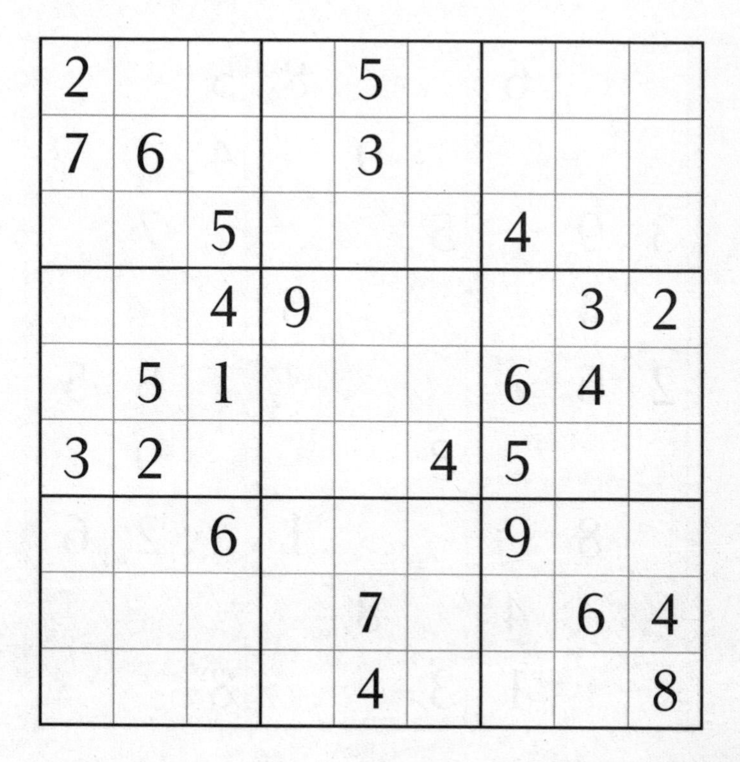

2				5				
7	6			3				
		5				4		
		4	9				3	2
	5	1				6	4	
3	2				4	5		
		6				9		
				7			6	4
				4				8

Fiendish

34

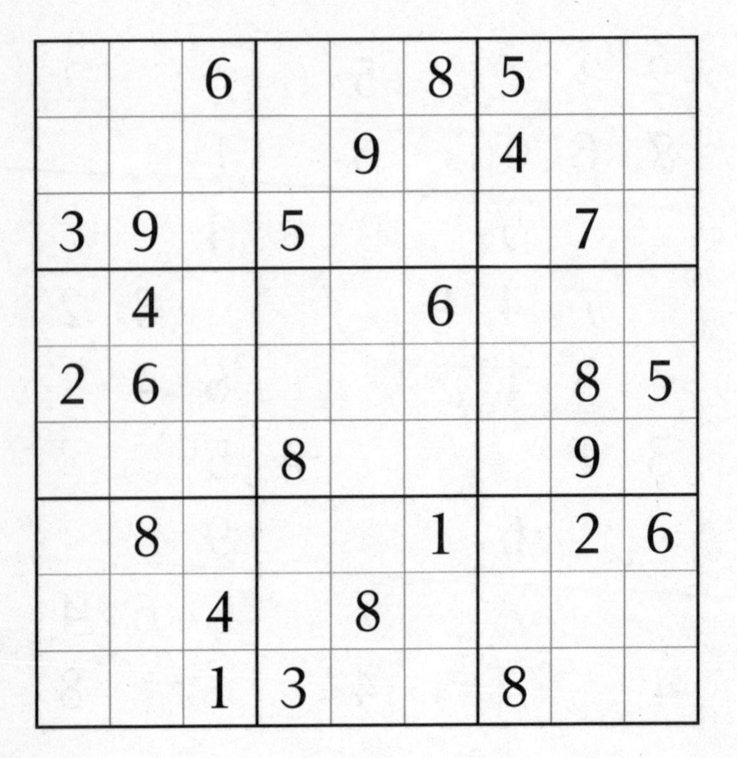

		6			8	5		
				9		4		
3	9		5				7	
	4				6			
2	6						8	5
			8				9	
	8				1		2	6
		4		8				
		1	3			8		

Fiendish

35

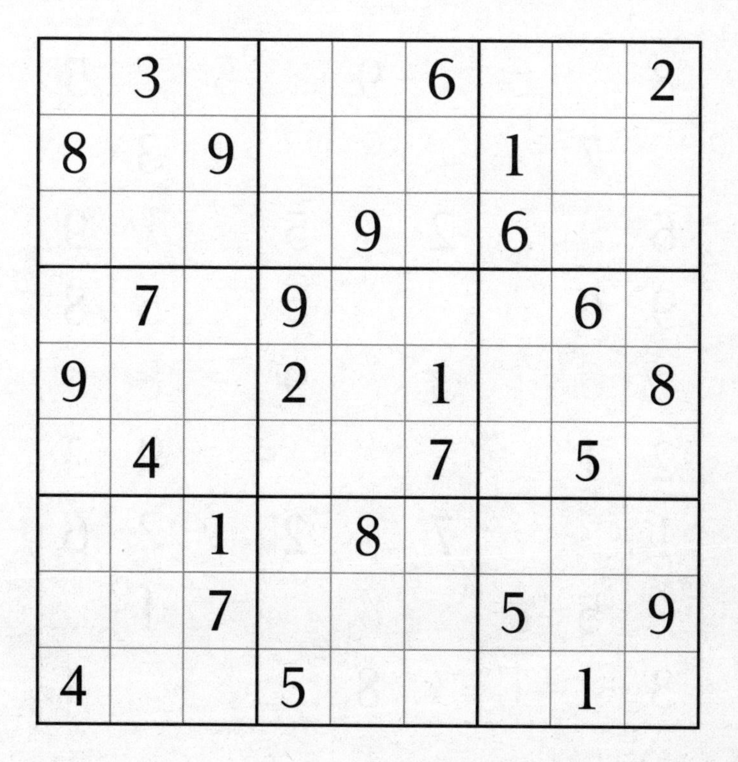

	3				6			2
8		9				1		
				9		6		
	7		9				6	
9			2		1			8
	4				7		5	
		1		8				
		7				5		9
4			5				1	

Fiendish

36

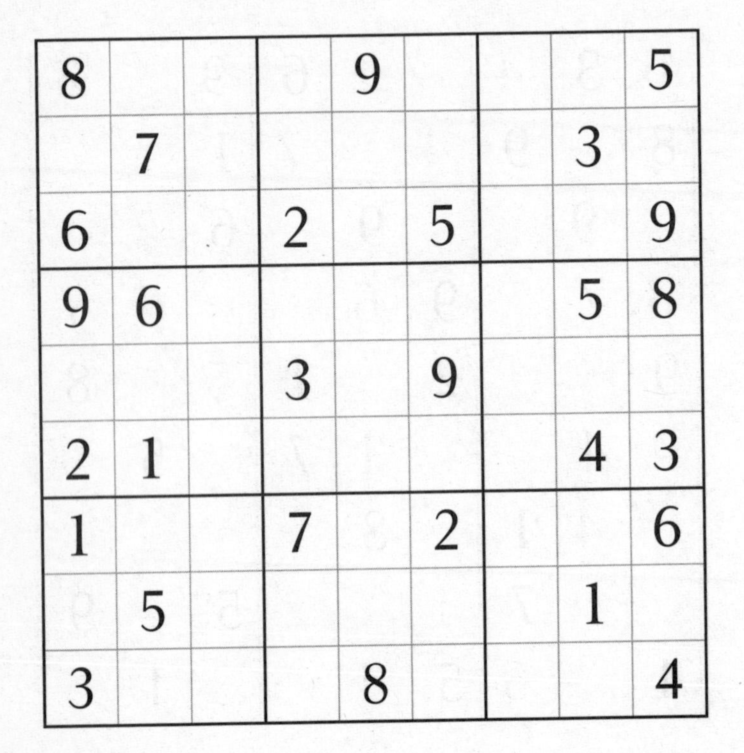

8				9				5
	7						3	
6			2		5			9
9	6						5	8
			3		9			
2	1						4	3
1			7		2			6
	5						1	
3				8				4

Fiendish

37

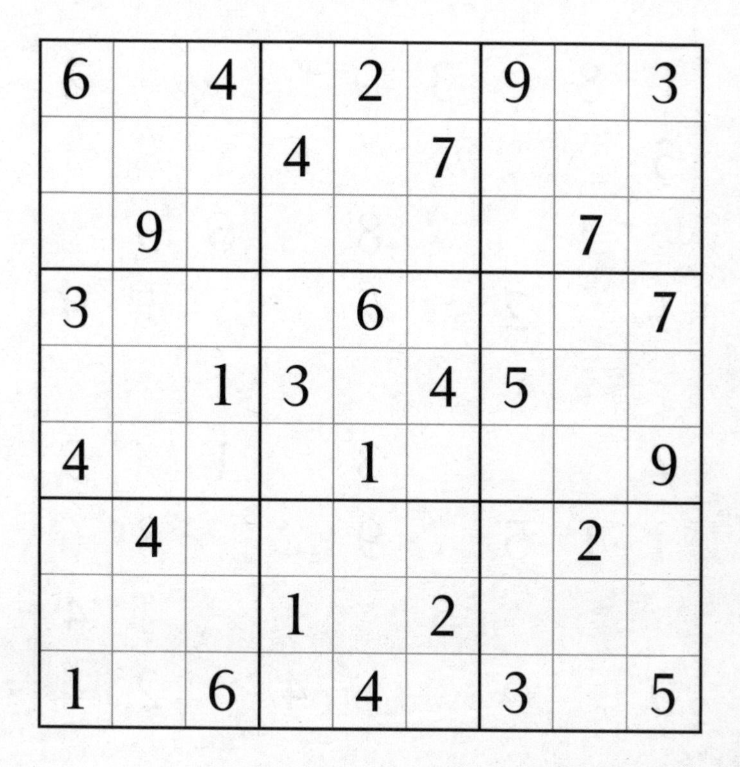

6		4		2		9		3
			4		7			
	9						7	
3				6				7
		1	3		4	5		
4				1				9
	4						2	
			1		2			
1		6		4		3		5

Fiendish

38

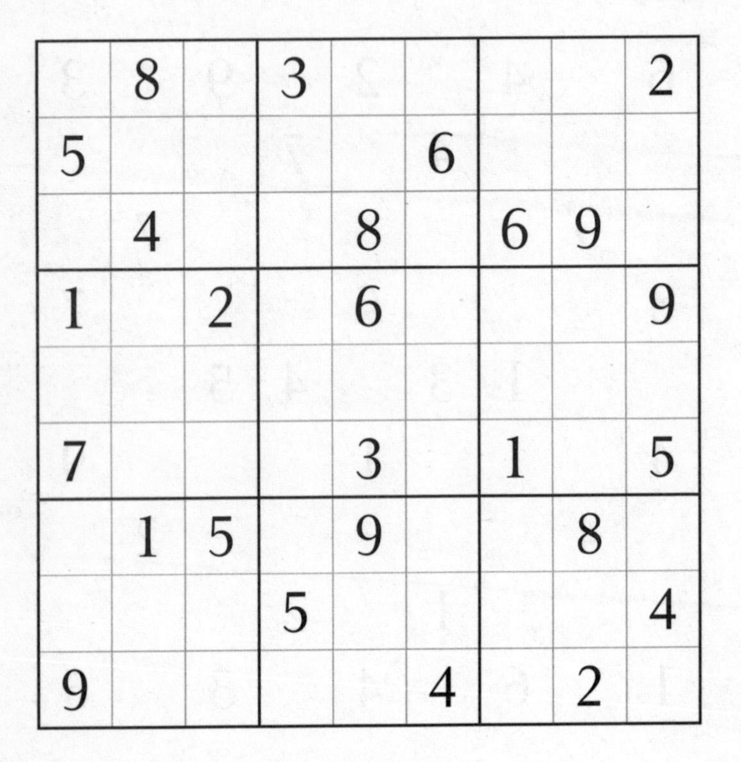

	8		3					2
5					6			
	4			8		6	9	
1		2		6				9
7				3		1		5
	1	5		9			8	
			5					4
9					4		2	

Fiendish

39

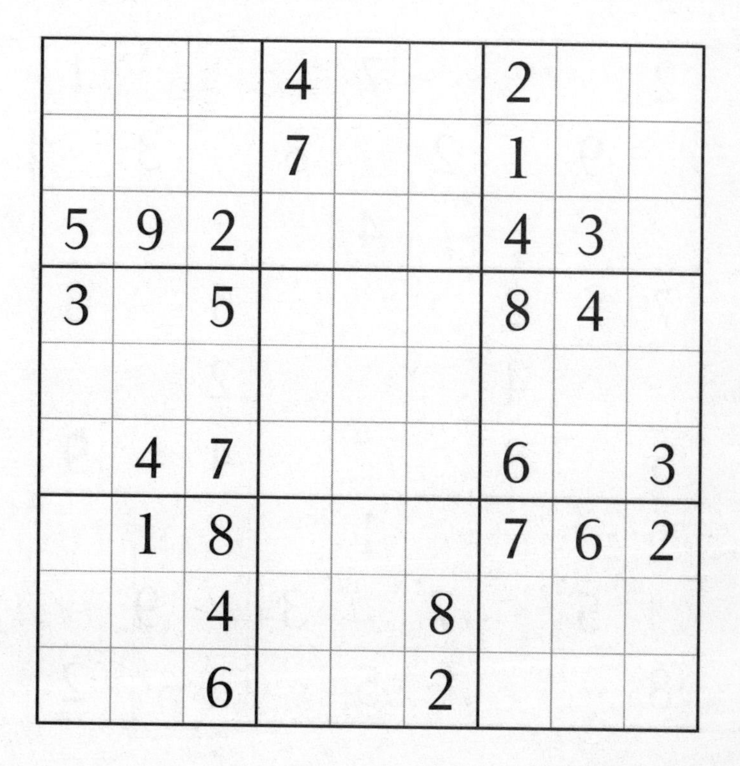

			4			2		
			7			1		
5	9	2				4	3	
3		5				8	4	
	4	7				6		3
	1	8				7	6	2
		4			8			
		6			2			

Fiendish

40

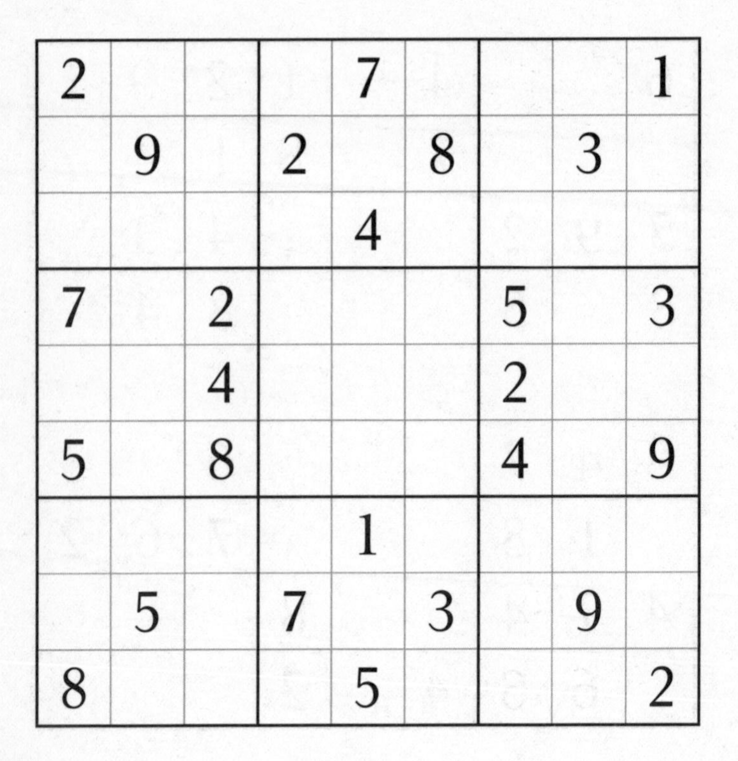

2				7				1
	9		2		8		3	
				4				
7		2				5		3
		4				2		
5		8				4		9
				1				
	5		7		3		9	
8				5				2

Fiendish

6					1	8	9	
					8		4	7
7								
		5	3					8
	4		8		2		1	
8					5	6		
								2
4	7		5					
	6	3	4					9

Fiendish

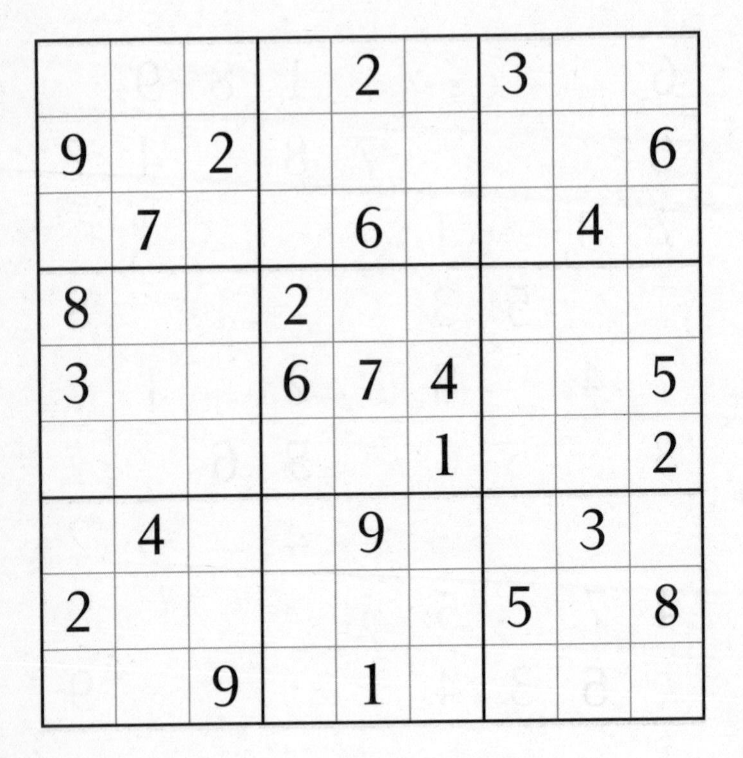

				2		3		
9		2						6
	7			6			4	
8			2					
3			6	7	4			5
					1			2
	4			9			3	
2						5		8
		9		1				

Fiendish

43

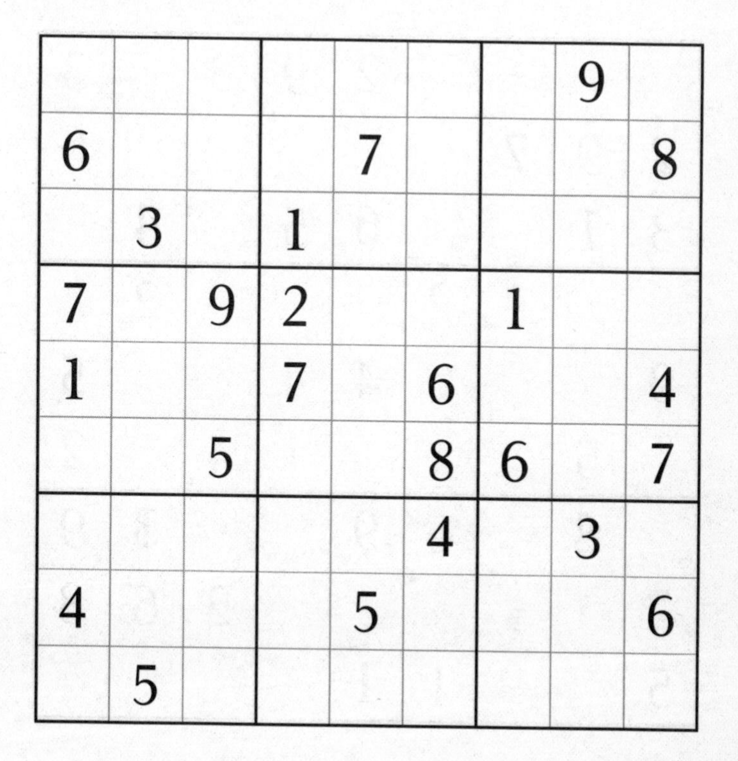

							9	
6				7				8
	3		1					
7		9	2			1		
1			7		6			4
		5			8	6		7
					4		3	
4				5				6
	5							

Fiendish

44

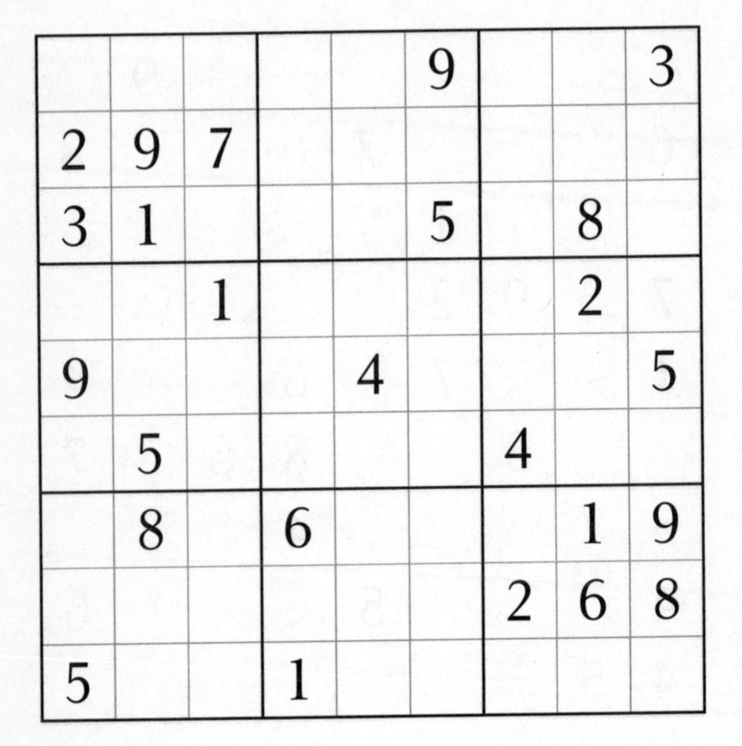

					9			3
2	9	7						
3	1				5		8	
		1					2	
9				4				5
	5					4		
	8		6				1	9
						2	6	8
5			1					

Fiendish

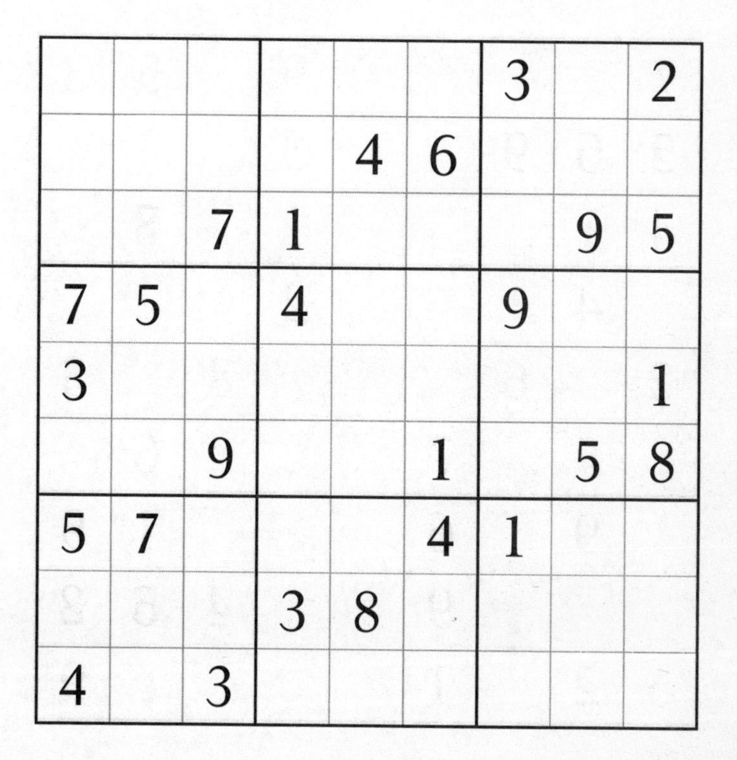

						3		2
				4	6			
		7	1				9	5
7	5		4			9		
3								1
		9			1		5	8
5	7				4	1		
			3	8				
4		3						

Fiendish

46

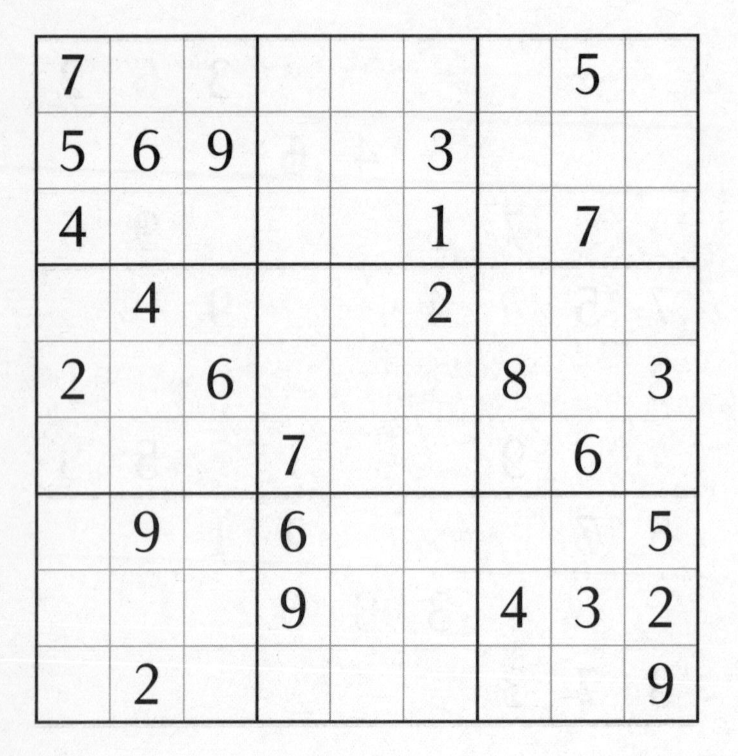

7							5	
5	6	9			3			
4					1		7	
	4				2			
2		6				8		3
			7				6	
	9		6					5
			9			4	3	2
	2							9

Fiendish

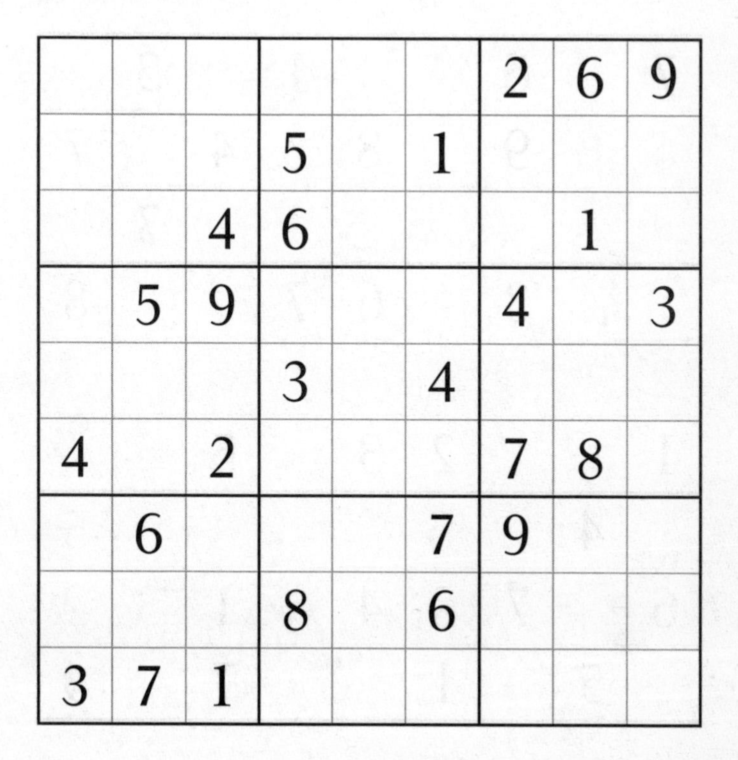

						2	6	9
			5		1			
		4	6				1	
	5	9				4		3
			3		4			
4		2				7	8	
	6				7	9		
			8		6			
3	7	1						

Fiendish

48

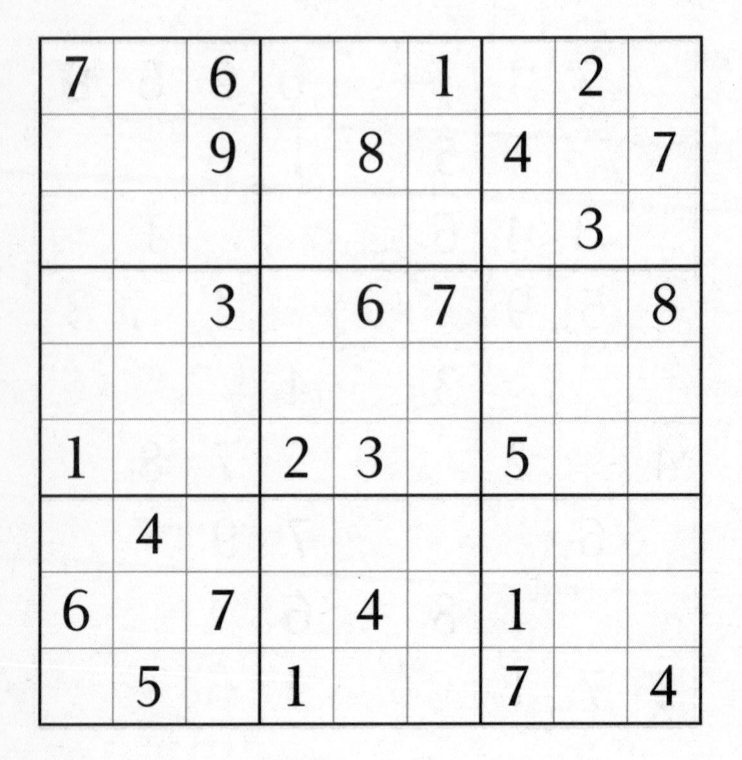

7		6			1		2	
		9		8		4		7
							3	
		3		6	7			8
1			2	3		5		
	4							
6		7		4		1		
	5		1			7		4

Fiendish

	7	1	5		6			9
		9	2			3		
	3							
			8	1				
9	6						3	8
				6	2			
							7	
		3			5	4		
7			3		1	5	9	

Fiendish

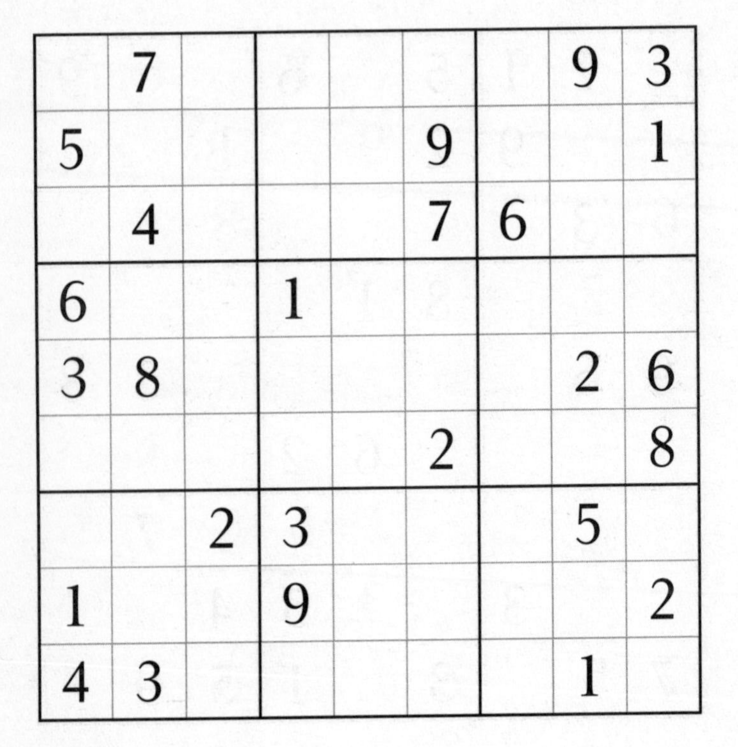

	7						9	3
5					9			1
	4				7	6		
6			1					
3	8						2	6
					2			8
		2	3				5	
1			9					2
4	3						1	

Fiendish

		5			8		6	3
				9		1		
6		7	5			8		
	7							
8			6		1			4
							1	
		3			5	7		1
		4		3				
7	5		8			6		

Fiendish

52

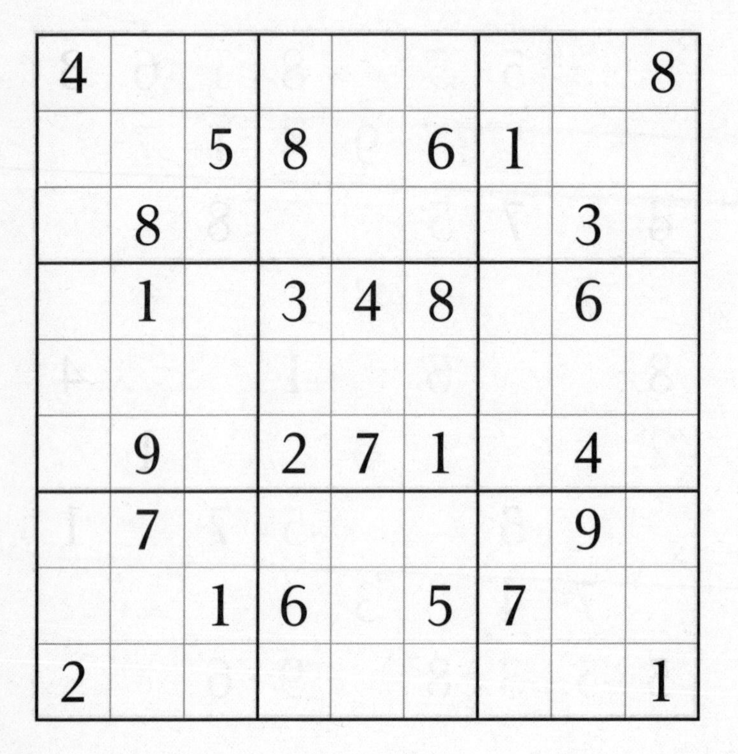

4								8
		5	8		6	1		
	8						3	
	1		3	4	8		6	
	9		2	7	1		4	
	7						9	
		1	6		5	7		
2								1

Fiendish

53

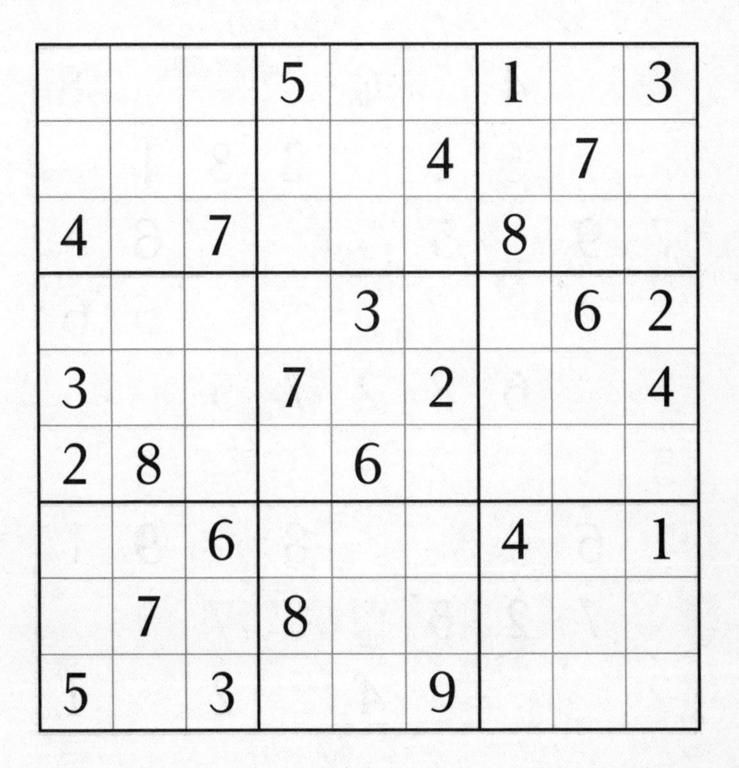

			5			1		3
					4		7	
4		7				8		
				3			6	2
3			7		2			4
2	8			6				
		6				4		1
	7		8					
5		3			9			

Fiendish

54

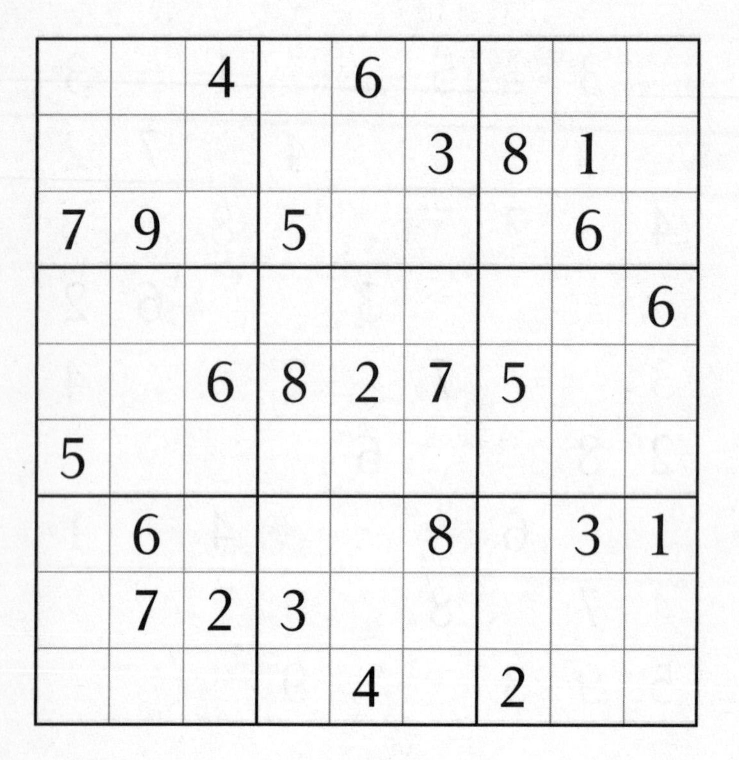

		4		6				
					3	8	1	
7	9		5				6	
								6
		6	8	2	7	5		
5								
	6				8		3	1
	7	2	3					
				4		2		

Fiendish

55

	3	2					6	
		8	4					2
		9	5					
		7		1				5
			9	3	4			
3				5		1		
					8	5		
4					9	7		
	9					6	8	

Fiendish

56

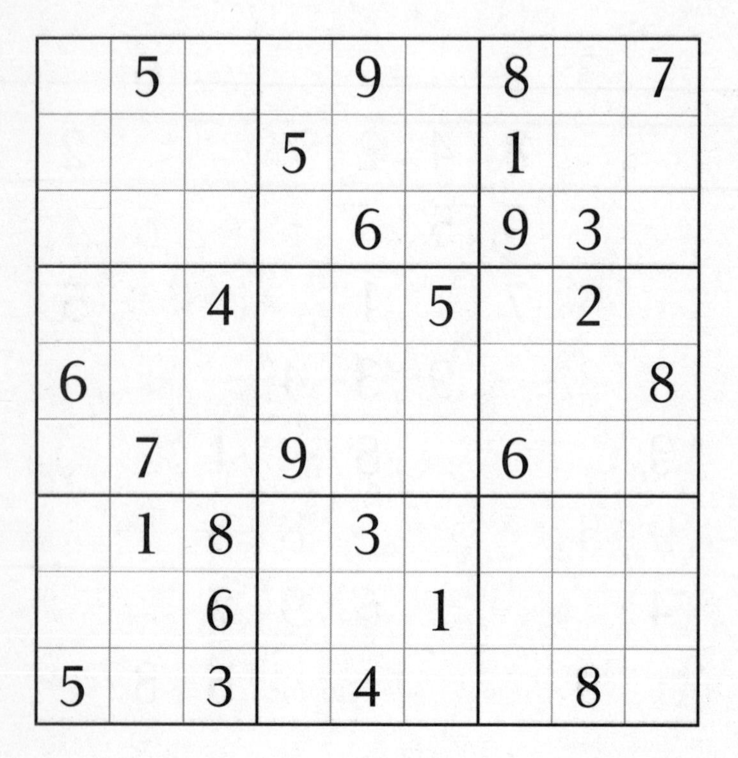

	5			9		8		7
			5			1		
				6		9	3	
		4			5		2	
6								8
	7		9			6		
	1	8		3				
		6			1			
5		3		4			8	

Fiendish

		8						
		3		2	7			
					1		6	4
	6	7		1				
	8		7	4	9		1	
				8		4	2	
9	5		4					
			9	6		8		
						5		

Fiendish

		8		6		2		
1			9				6	
5			4					
2						3	9	
4			6		2			8
	7	1						2
					7			9
	5				4			6
		4		8		7		

Fiendish

59

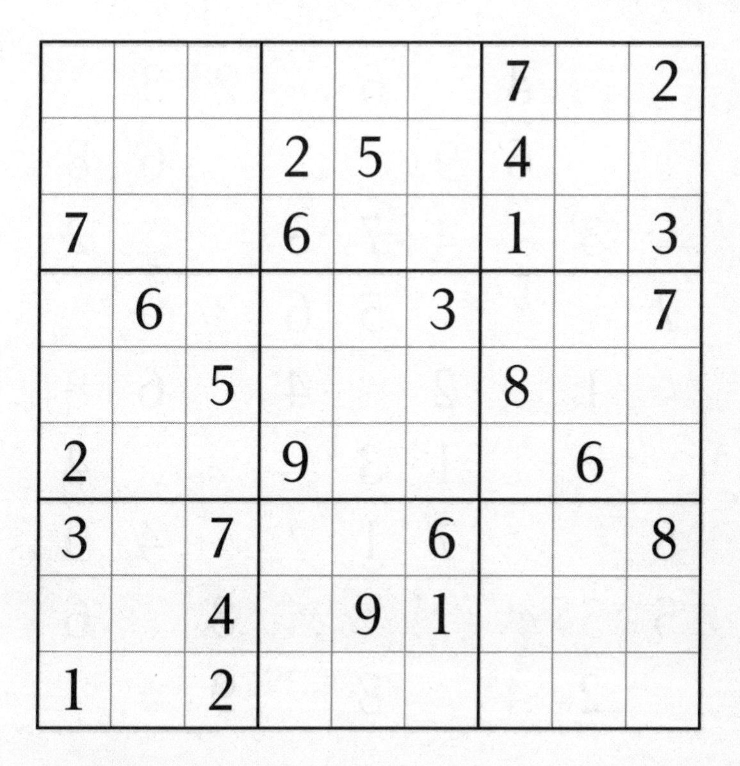

						7		2
			2	5		4		
7			6			1		3
	6				3			7
		5				8		
2			9				6	
3		7			6			8
		4		9	1			
1		2						

Fiendish

60

		4					3	
		6						8
	3			7	8			2
7				5	6			
	1		2		4		6	
			1	3				4
8			7	1			4	
5						8		
	2					9		

Fiendish

61

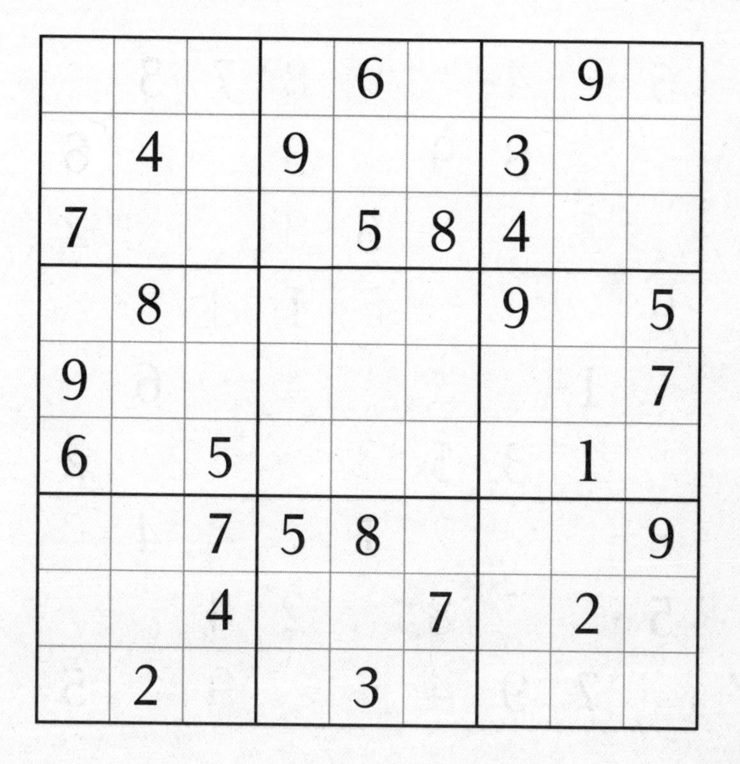

				6			9	
	4		9			3		
7				5	8	4		
	8					9		5
9								7
6		5					1	
		7	5	8				9
		4			7		2	
	2			3				

Fiendish

62

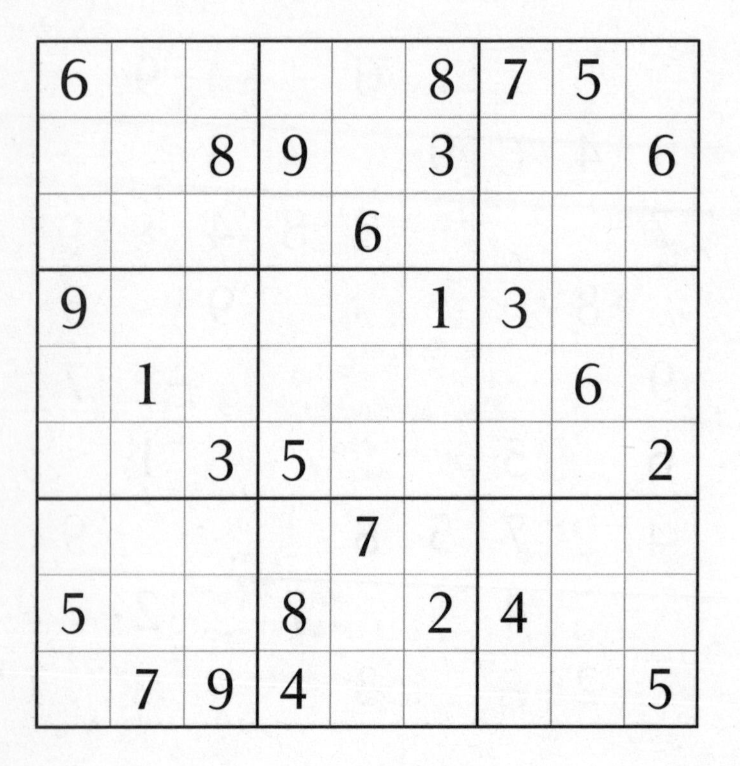

6					8	7	5	
		8	9		3			6
				6				
9					1	3		
	1						6	
		3	5					2
				7				
5			8		2	4		
	7	9	4					5

Fiendish

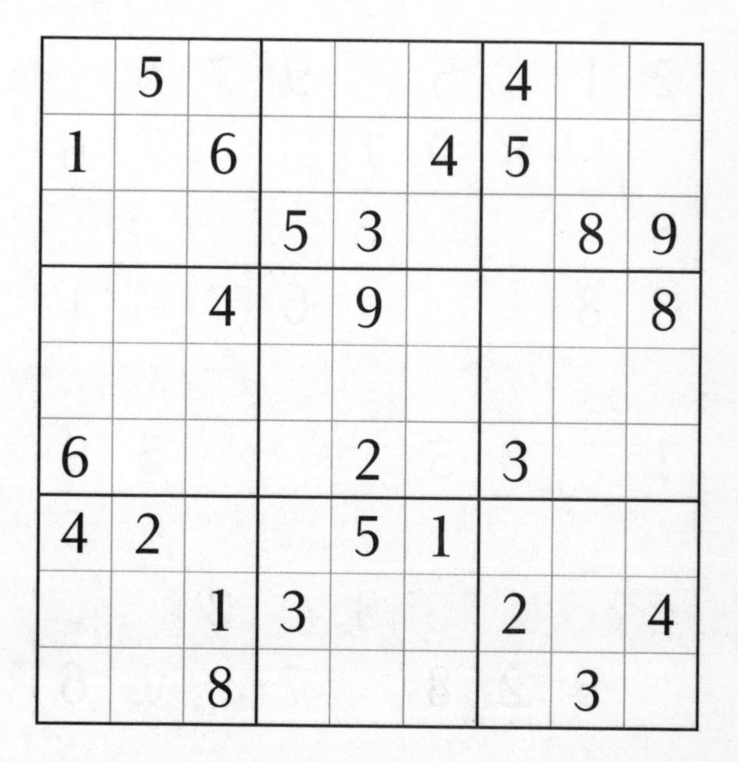

	5					4		
1		6			4	5		
			5	3			8	9
		4		9				8
6				2		3		
4	2			5	1			
		1	3			2		4
		8					3	

Fiendish

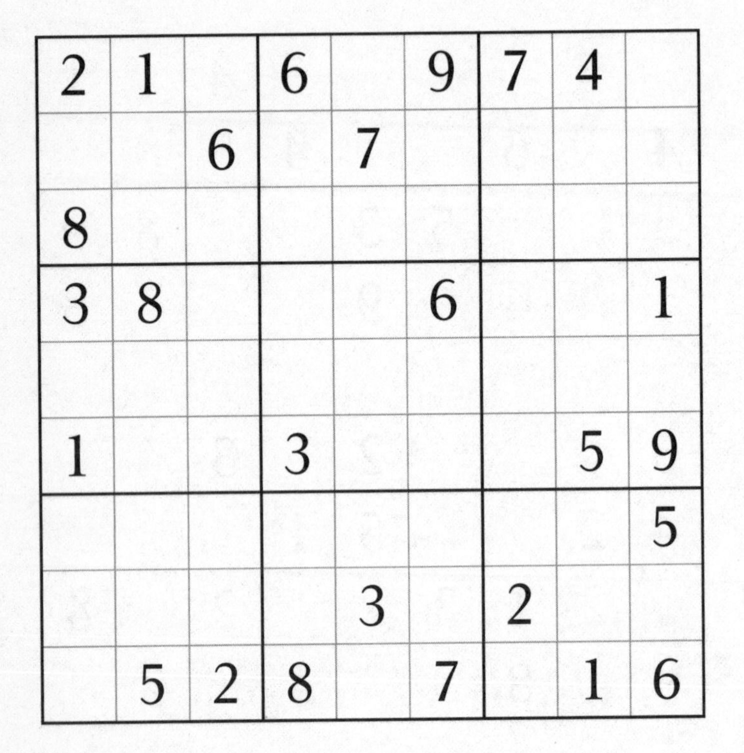

2	1		6		9	7	4	
		6		7				
8								
3	8				6			1
1			3				5	9
								5
				3		2		
	5	2	8		7		1	6

Fiendish

65

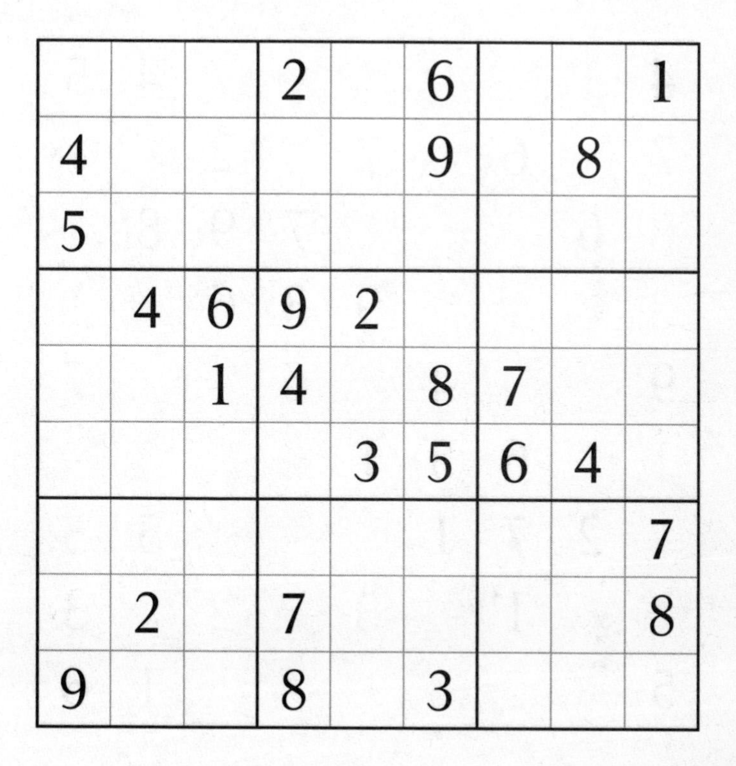

			2		6			1
4					9		8	
5								
	4	6	9	2				
		1	4		8	7		
				3	5	6	4	
								7
	2		7					8
9			8		3			

Fiendish

66

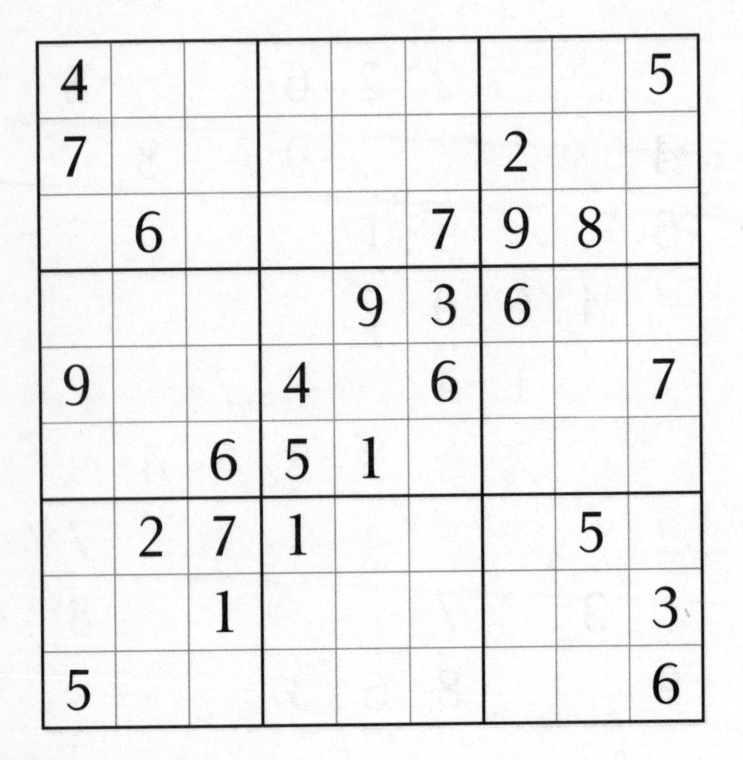

4								5
7						2		
	6				7	9	8	
				9	3	6		
9			4		6			7
		6	5	1				
	2	7	1				5	
		1						3
5								6

Fiendish

67

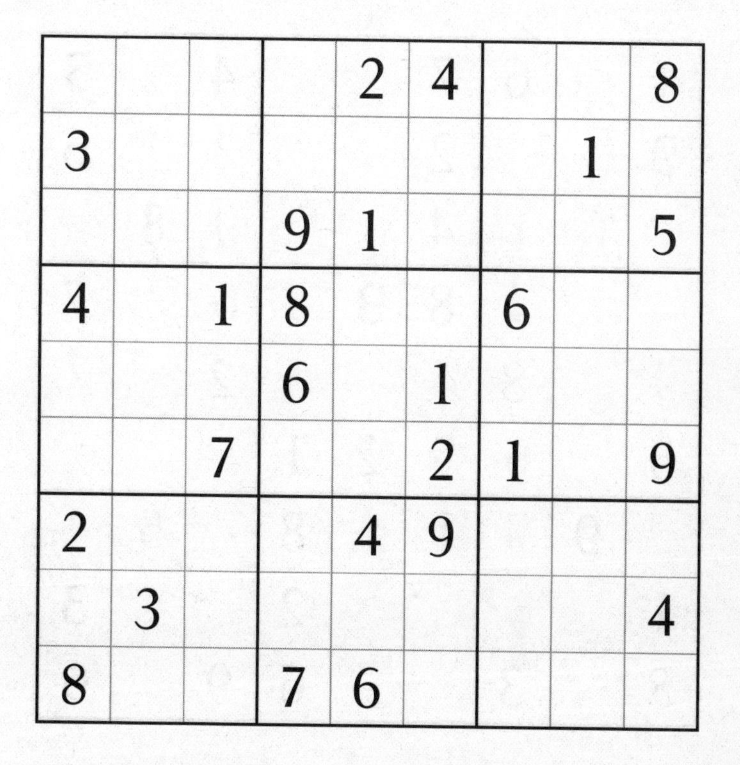

				2	4			8
3							1	
			9	1				5
4		1	8			6		
			6		1			
		7			2	1		9
2				4	9			
	3							4
8			7	6				

Fiendish

		6	7			4		3
9			2					6
			4				1	
			8	3				7
		8				2		
5				2	1			
	9				8			
1					2			5
8		3			6	9		

Fiendish

69

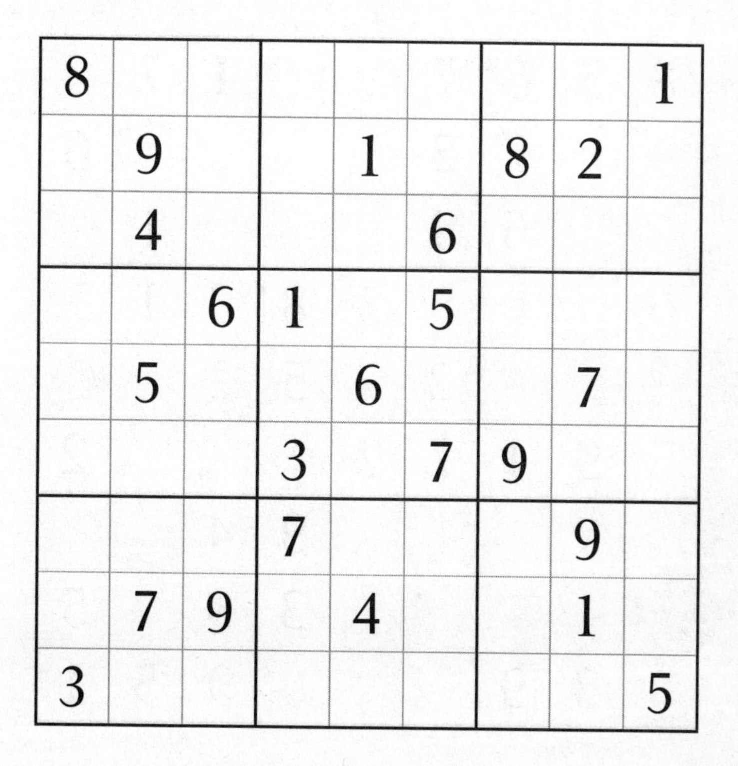

8								1
	9			1		8	2	
	4				6			
		6	1		5			
	5			6			7	
			3		7	9		
			7				9	
	7	9		4			1	
3								5

Fiendish

70

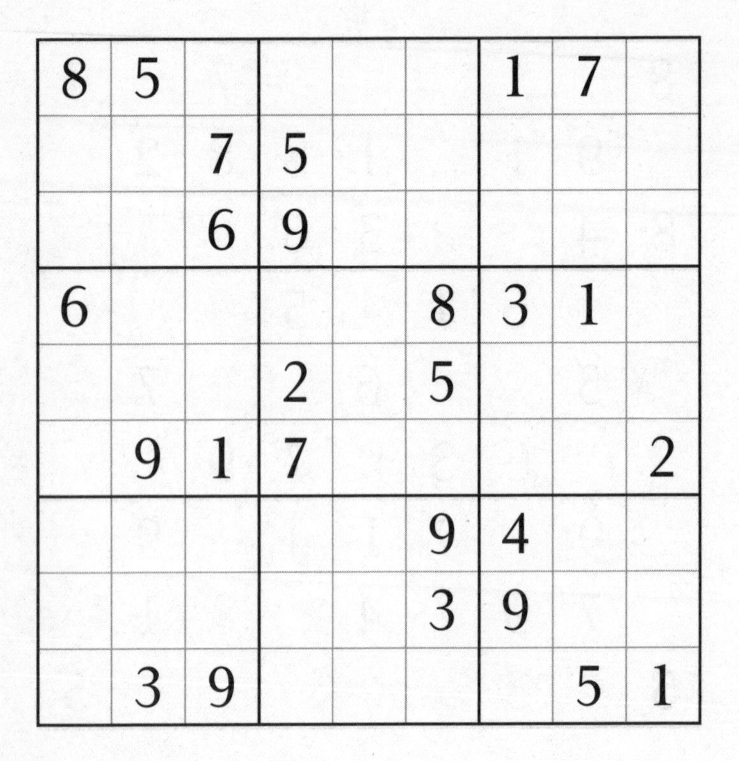

8	5					1	7	
		7	5					
		6	9					
6					8	3	1	
			2		5			
	9	1	7					2
					9	4		
					3	9		
	3	9					5	1

Fiendish

71

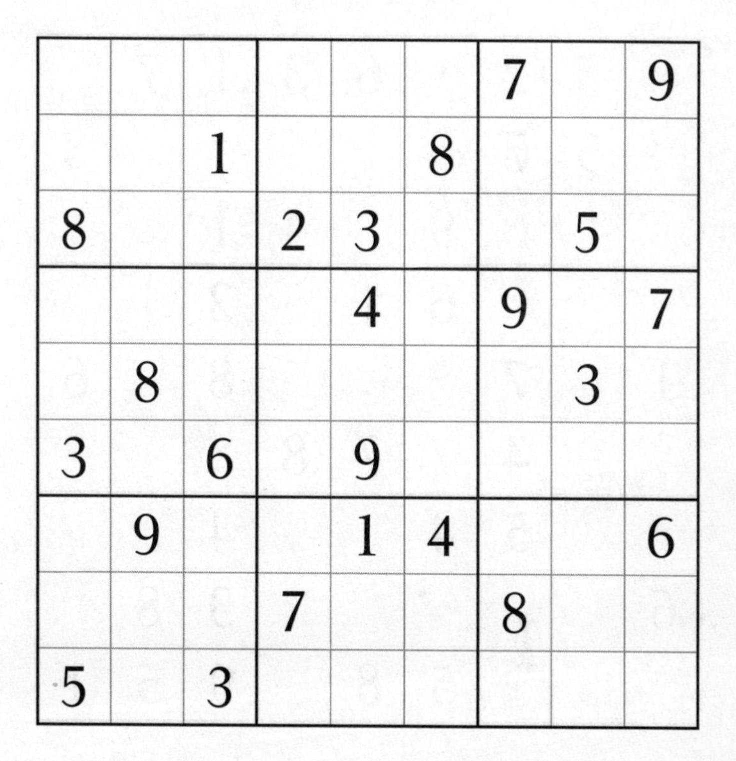

						7		9
		1			8			
8			2	3			5	
				4		9		7
	8						3	
3		6		9				
	9			1	4			6
			7			8		
5		3						

Fiendish

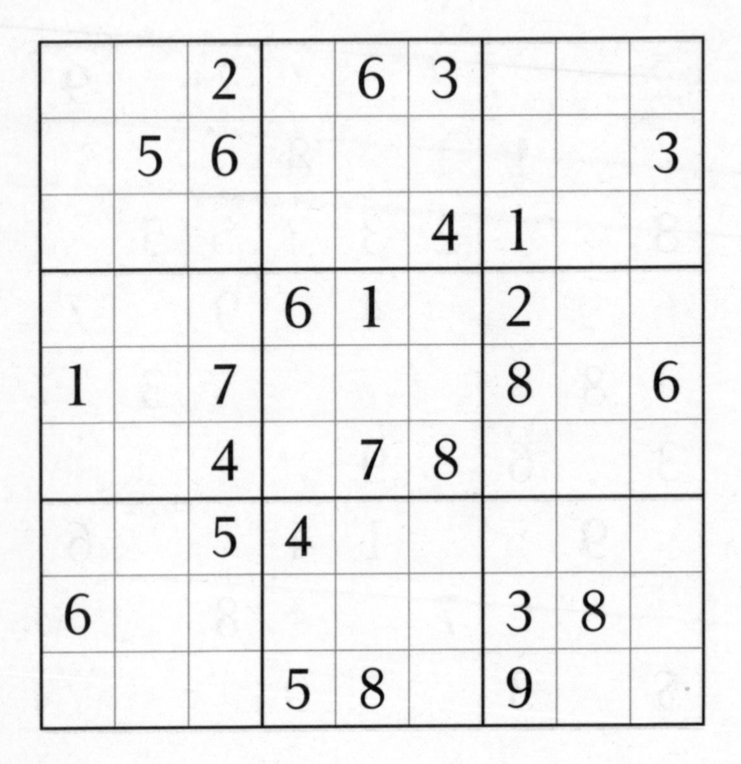

		2		6	3			
	5	6						3
					4	1		
			6	1		2		
1		7				8		6
		4		7	8			
		5	4					
6						3	8	
			5	8		9		

Fiendish

73

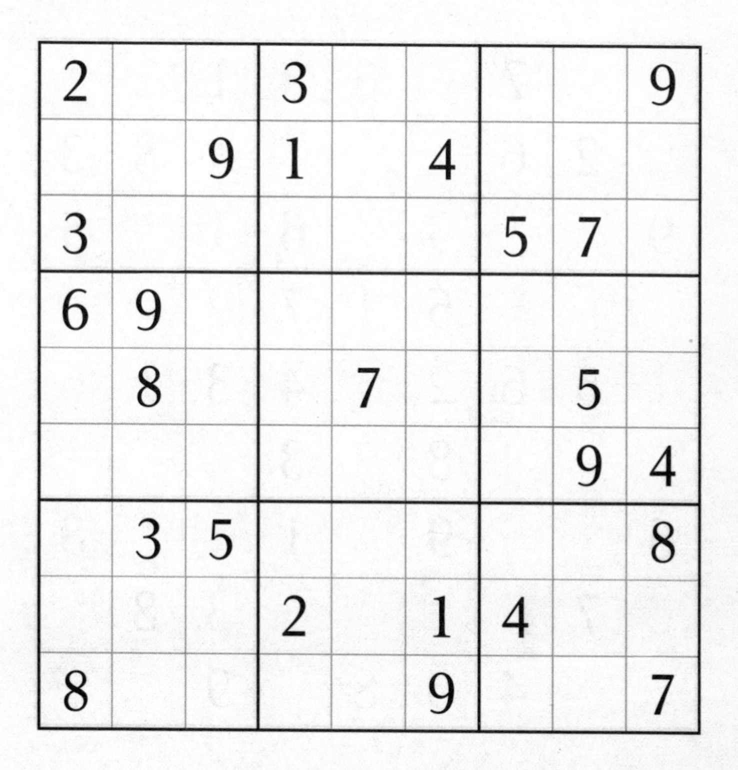

2			3					9
		9	1		4			
3						5	7	
6	9							
	8			7			5	
							9	4
	3	5						8
			2		1	4		
8					9			7

Fiendish

		7				1		
	2						8	
9			3		6			5
			5		7			
	5	6	2		4	3	1	
			8		3			
6			9		1			8
	7						2	
		4				5		

Fiendish

		4		2				3
	1							9
8			4	5			7	
			3		5		6	
	3		7		1			
	4			3	2			5
9							3	
7				4		6		

Fiendish

76

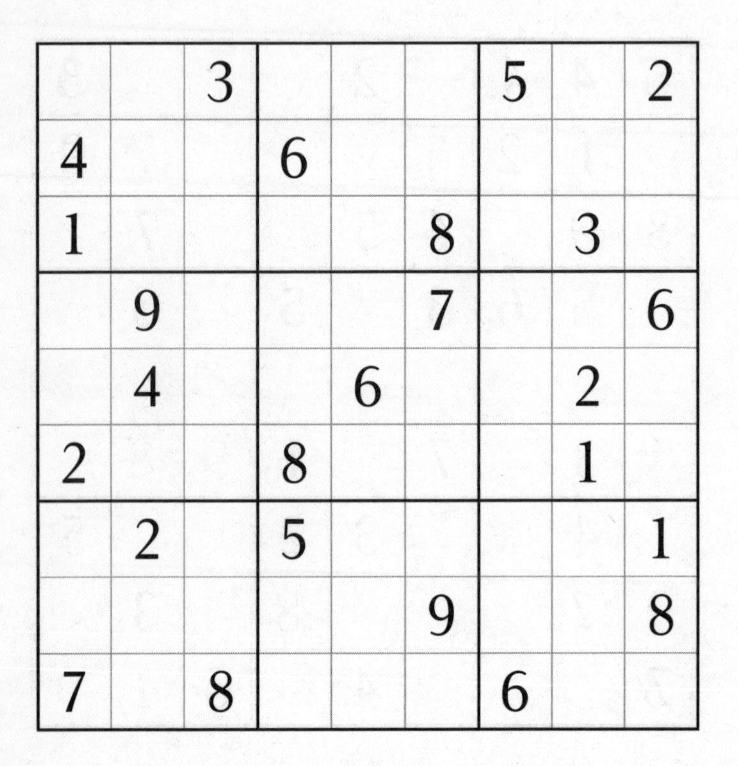

		3				5		2
4			6					
1					8		3	
	9				7			6
	4			6			2	
2			8				1	
	2		5					1
					9			8
7		8				6		

Fiendish

77

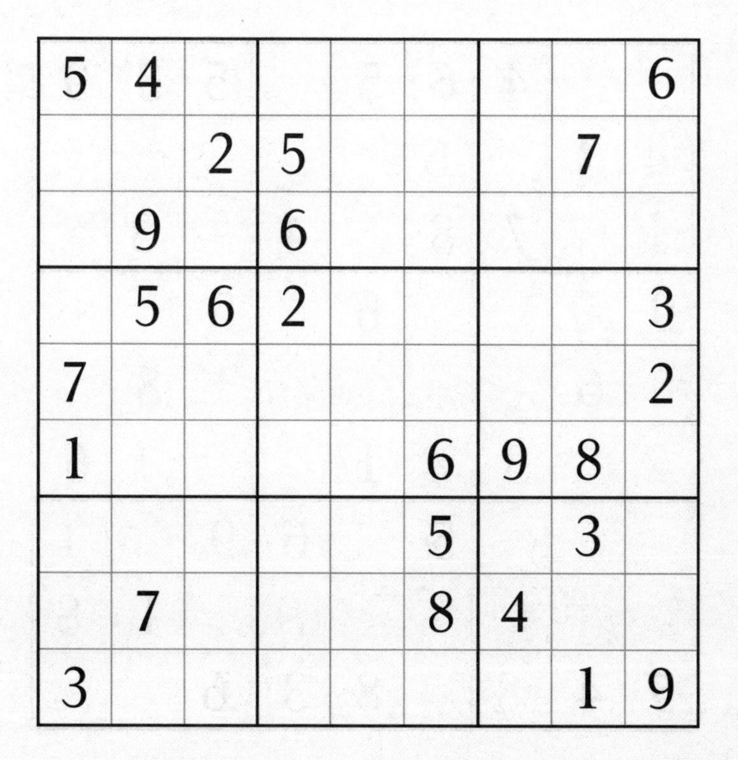

5	4							6
		2	5				7	
	9		6					
	5	6	2					3
7								2
1					6	9	8	
					5		3	
	7				8	4		
3							1	9

Fiendish

78

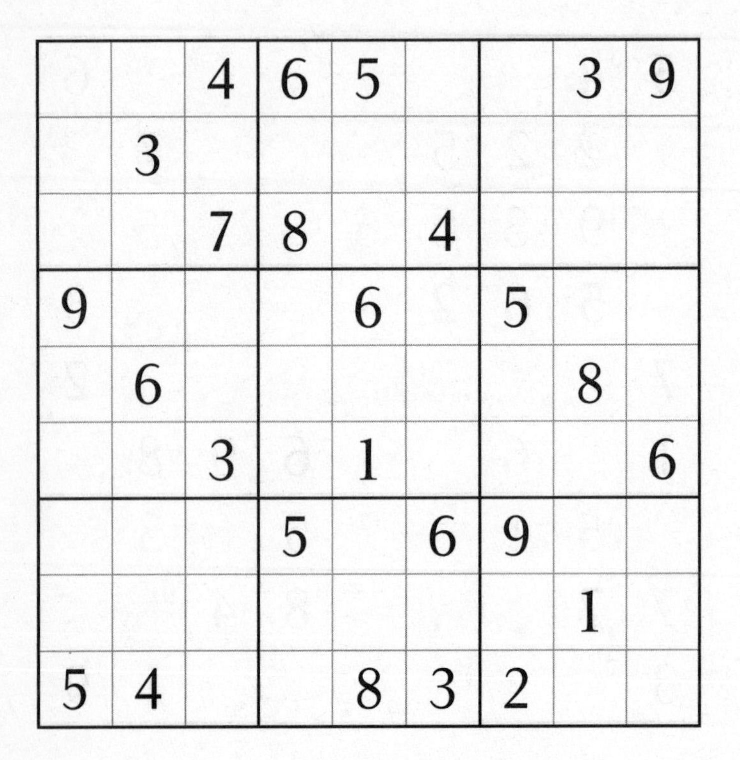

		4	6	5			3	9
	3							
		7	8		4			
9				6		5		
	6						8	
		3		1				6
			5		6	9		
							1	
5	4			8	3	2		

Fiendish

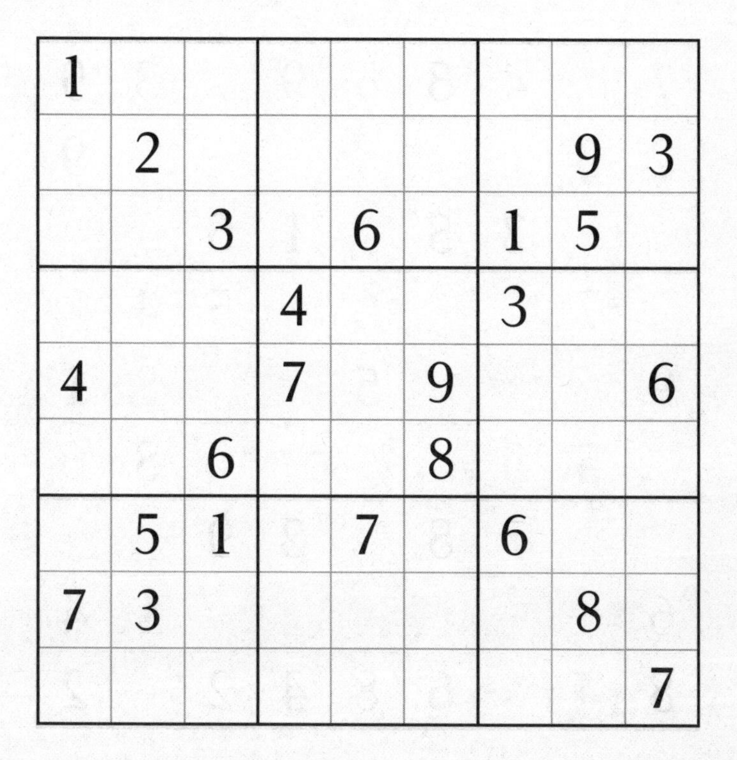

1								
	2						9	3
		3		6		1	5	
			4			3		
4			7		9			6
		6			8			
	5	1		7		6		
7	3						8	
								7

Fiendish

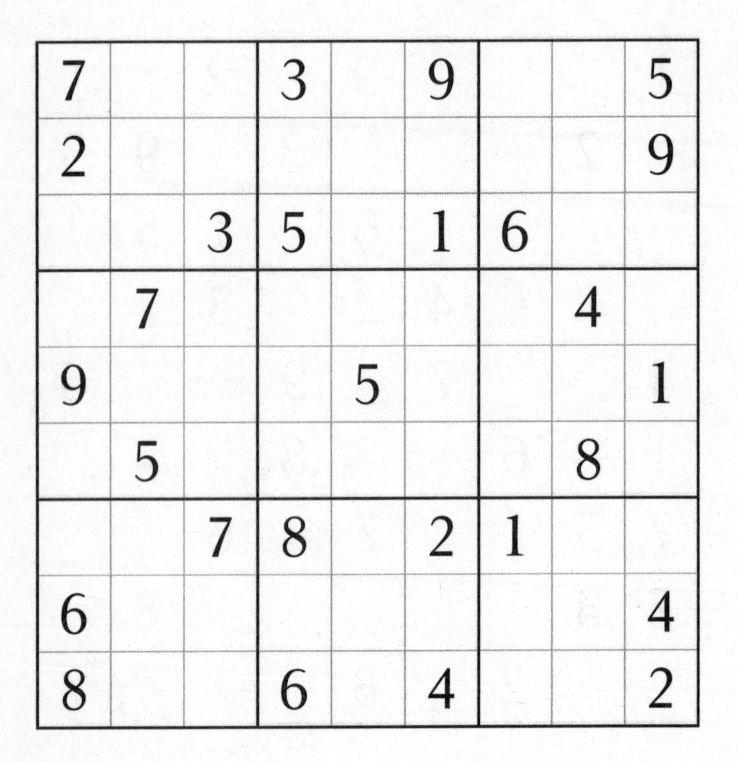

7			3		9			5
2								9
		3	5		1	6		
	7						4	
9				5				1
	5						8	
		7	8		2	1		
6								4
8			6		4			2

Fiendish

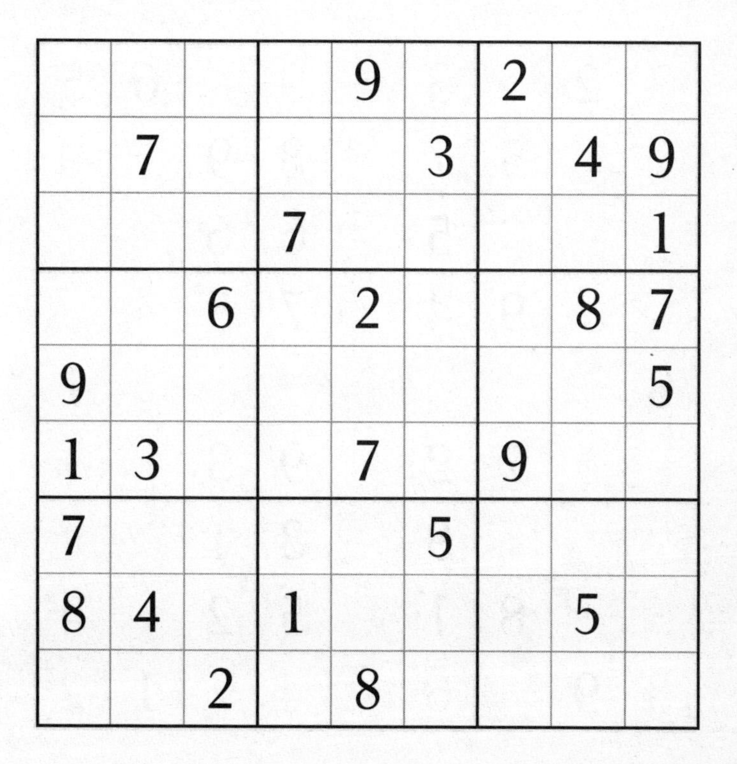

				9		2		
	7				3		4	9
			7					1
		6		2			8	7
9								5
1	3			7		9		
7					5			
8	4		1				5	
		2		8				

Fiendish

82

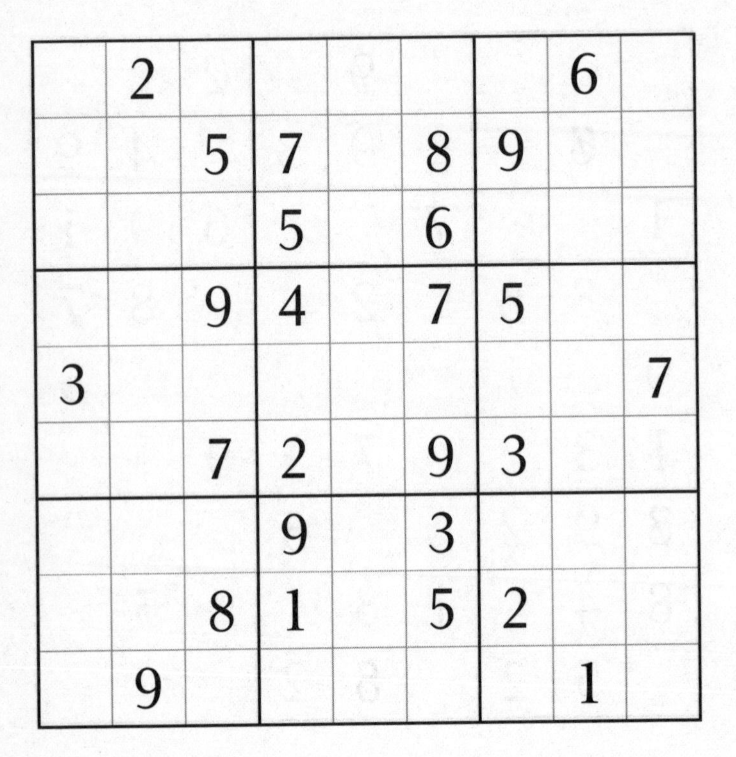

	2						6	
		5	7		8	9		
			5		6			
		9	4		7	5		
3								7
		7	2		9	3		
			9		3			
		8	1		5	2		
	9						1	

Fiendish

			1	7			2	
	6			9				
1						6	9	5
	8		9					1
		7				4		
3					1		5	
8	9	2						3
				8			7	
	5			6	2			

Fiendish

84

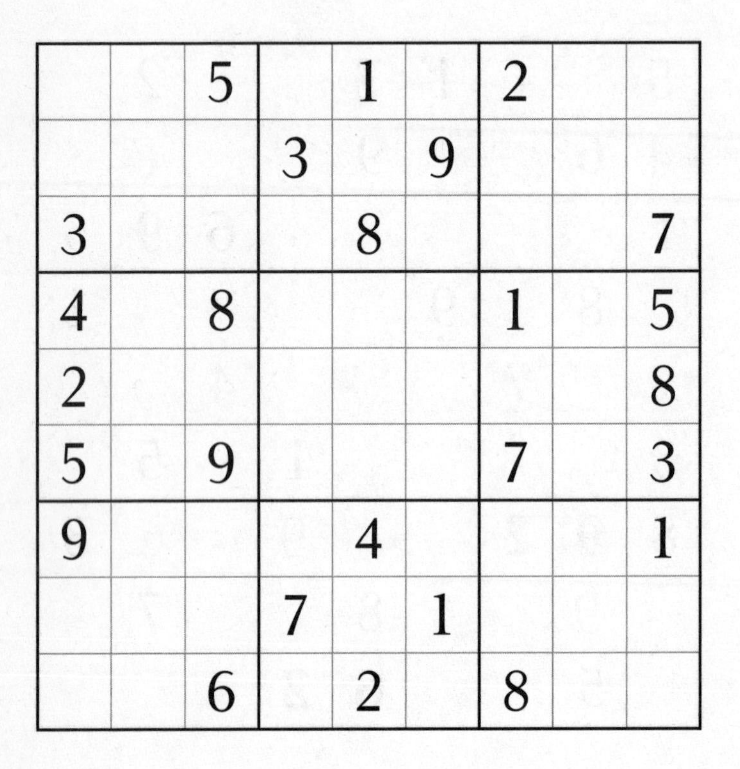

		5		1		2		
			3		9			
3				8				7
4		8				1		5
2								8
5		9				7		3
9				4				1
			7		1			
		6		2		8		

Fiendish

5		6	8	3				
4							8	
			4			6	1	
			9			5		
2				6				7
		8			7			
	6	5			9			
	9							4
				8	5	1		9

Fiendish

86

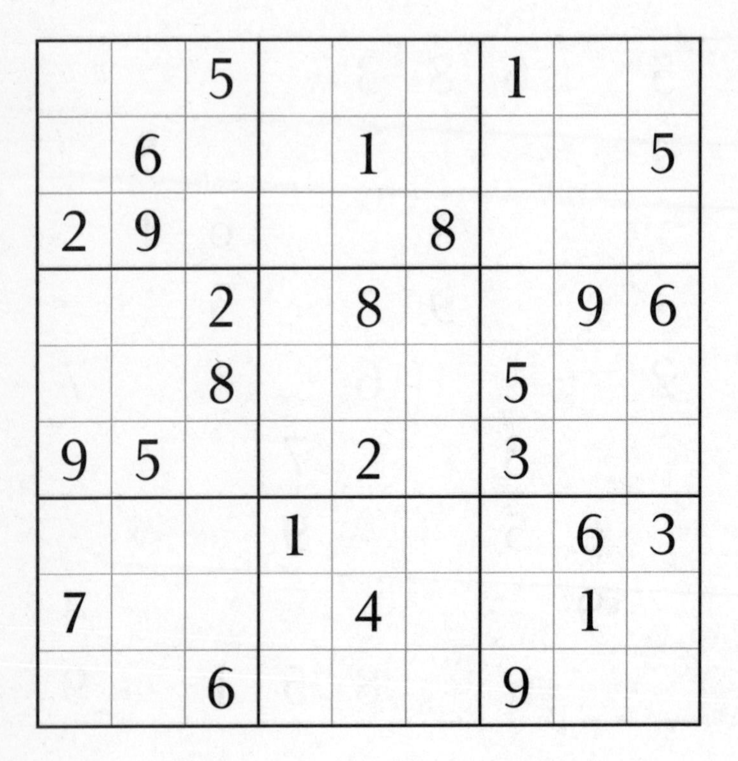

		5				1		
	6			1				5
2	9				8			
		2		8			9	6
		8				5		
9	5			2		3		
			1				6	3
7				4			1	
		6				9		

Fiendish

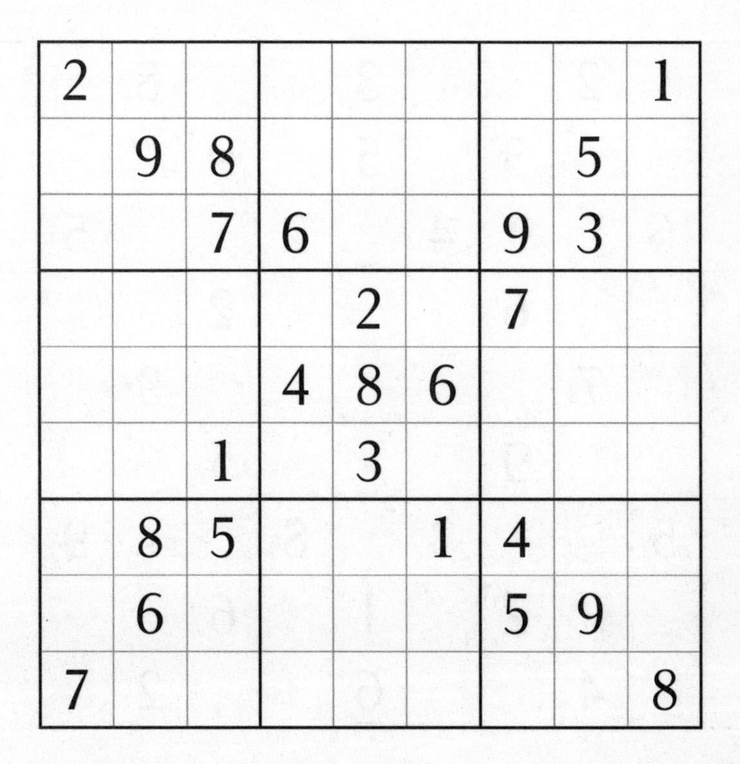

2								1
	9	8					5	
		7	6			9	3	
				2		7		
			4	8	6			
		1		3				
	8	5			1	4		
	6					5	9	
7								8

Fiendish

88

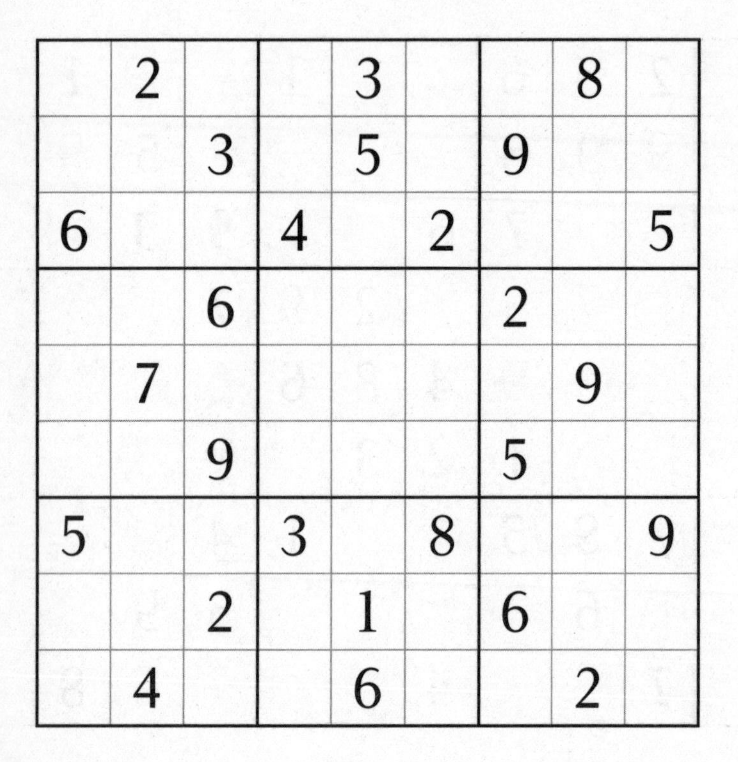

	2			3			8	
		3		5		9		
6			4		2			5
		6				2		
	7						9	
		9				5		
5			3		8			9
		2		1		6		
	4			6			2	

Fiendish

89

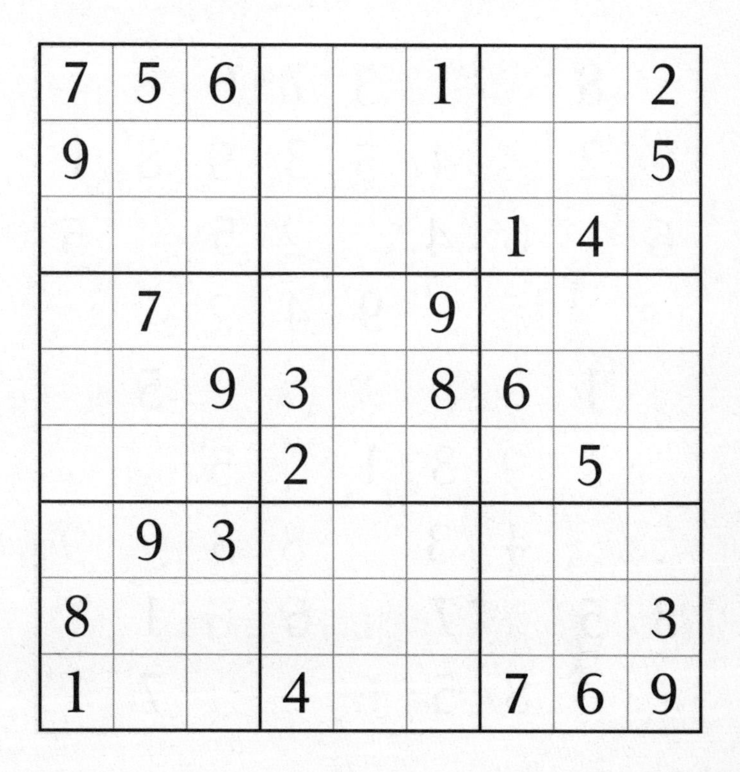

7	5	6			1			2
9								5
						1	4	
	7				9			
		9	3		8	6		
			2				5	
	9	3						
8								3
1			4			7	6	9

Fiendish

90

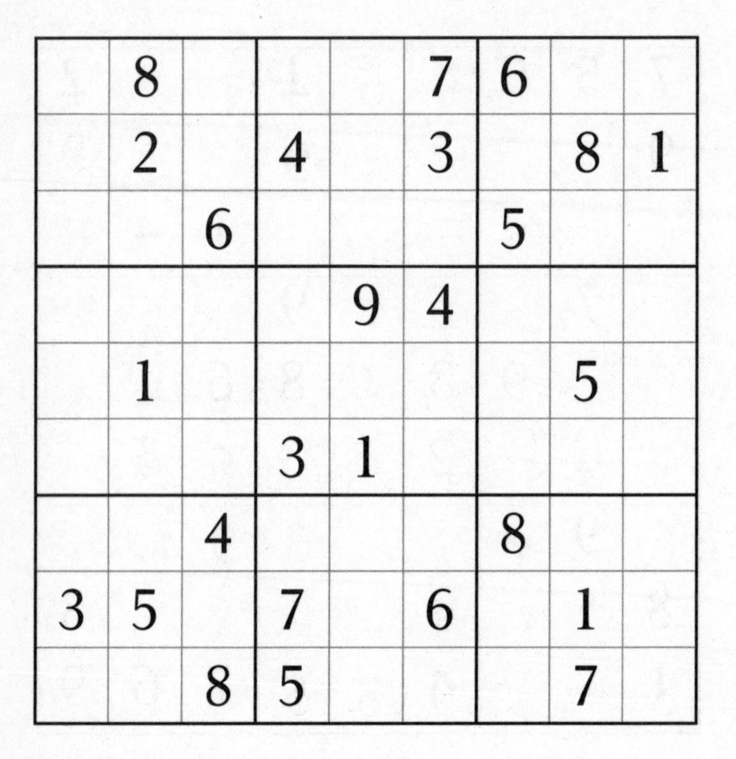

	8				7	6		
	2		4		3		8	1
		6				5		
				9	4			
	1						5	
			3	1				
		4				8		
3	5		7		6		1	
		8	5				7	

Fiendish

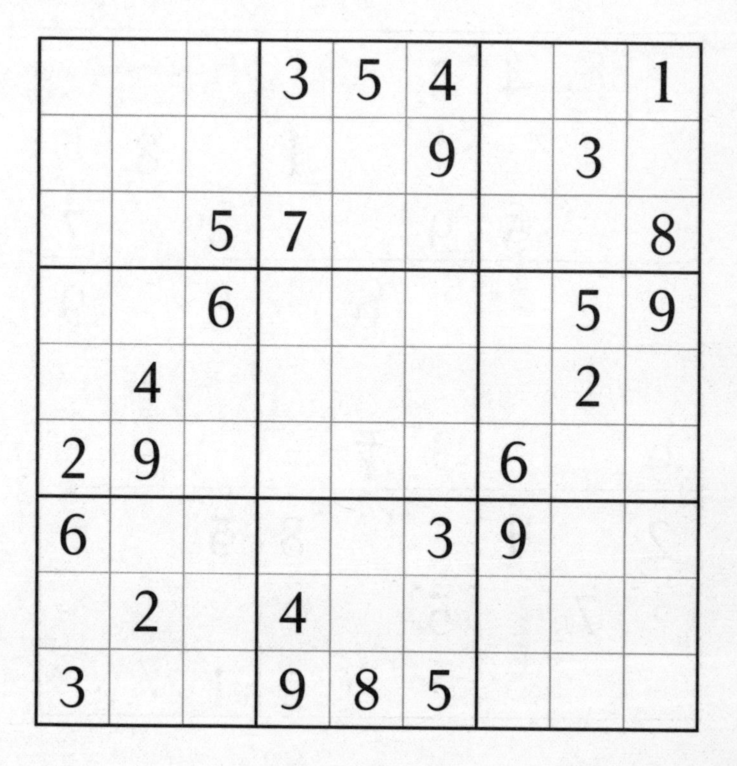

			3	5	4			1
					9		3	
		5	7					8
		6					5	9
	4						2	
2	9					6		
6					3	9		
	2		4					
3			9	8	5			

Fiendish

92

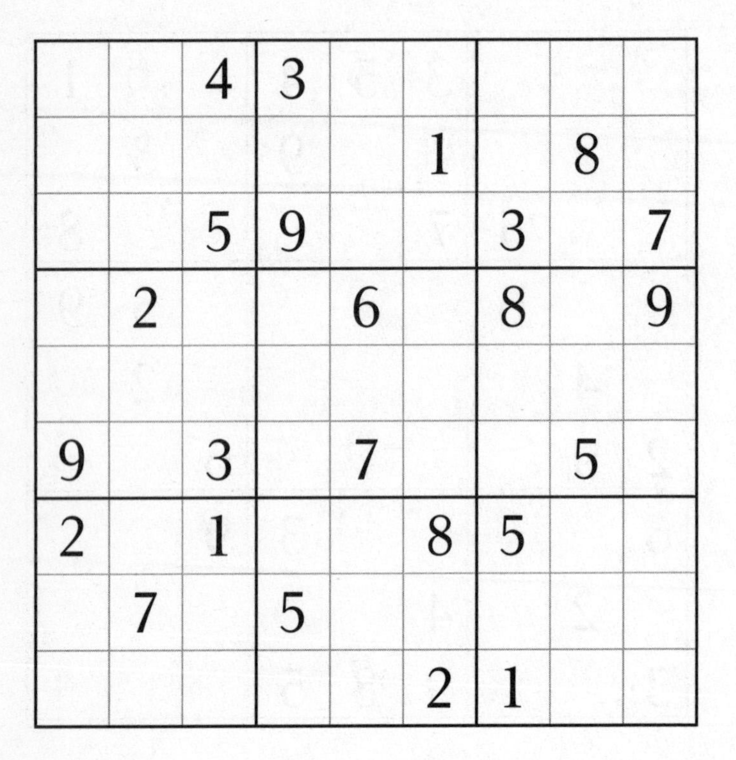

		4	3					
					1		8	
		5	9			3		7
	2			6		8		9
9		3		7			5	
2		1			8	5		
	7		5					
					2	1		

Fiendish

93

3				5			6	
	2		4				9	
		1				8		
6			1	2				
7								9
				9	4			3
		3				9		
	9				5		1	
	5			7				4

Fiendish

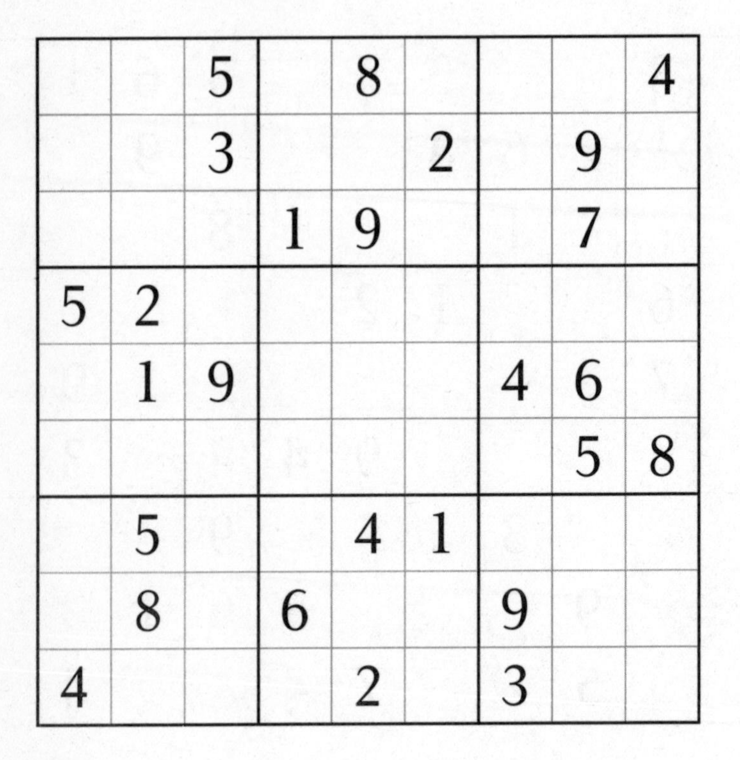

		5		8				4
		3			2		9	
			1	9			7	
5	2							
	1	9				4	6	
							5	8
	5			4	1			
	8		6			9		
4				2		3		

Fiendish

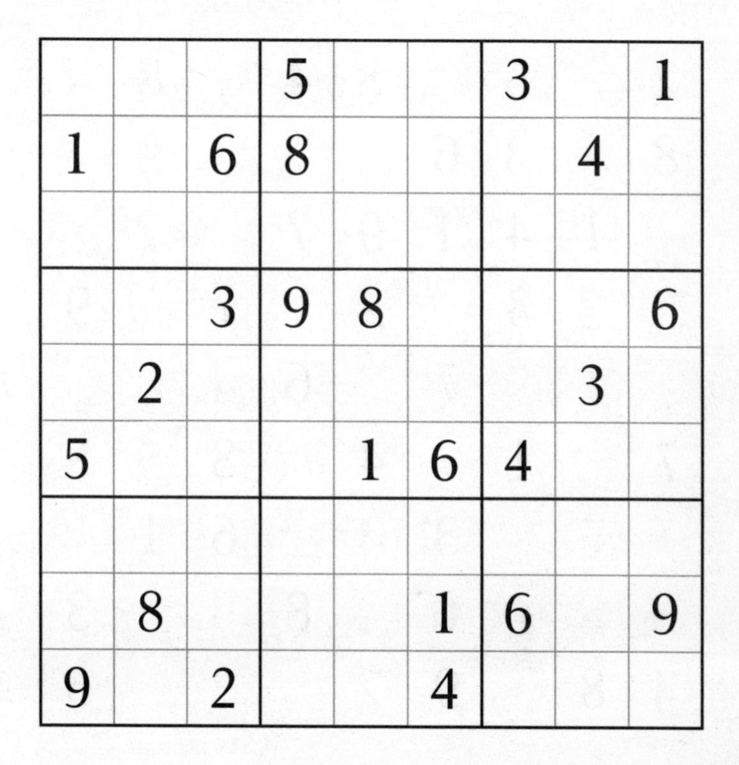

			5			3		1
1		6	8				4	
		3	9	8				6
	2						3	
5				1	6	4		
	8				1	6		9
9		2			4			

Fiendish

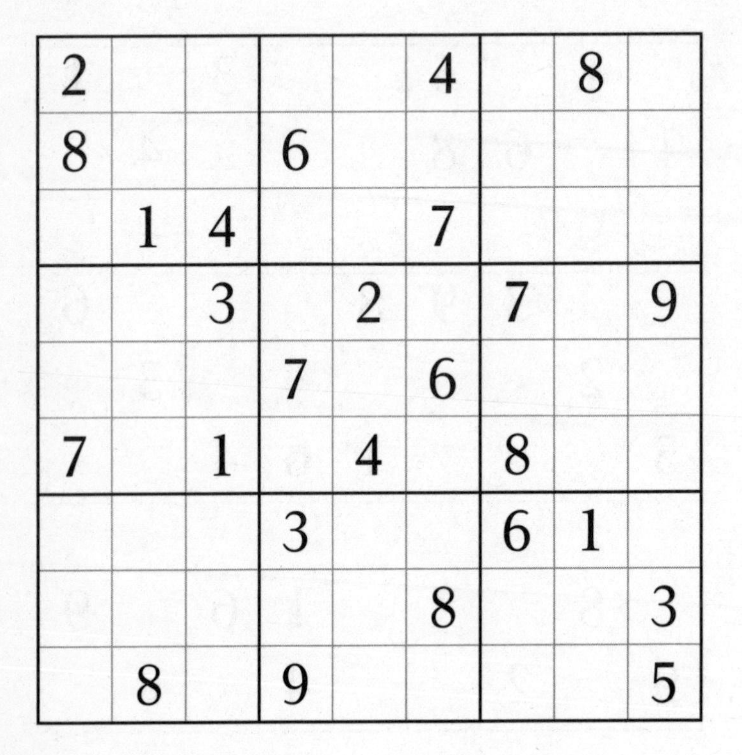

2					4		8	
8			6					
	1	4			7			
		3		2		7		9
			7		6			
7		1		4		8		
			3			6	1	
					8			3
	8		9					5

Fiendish

97

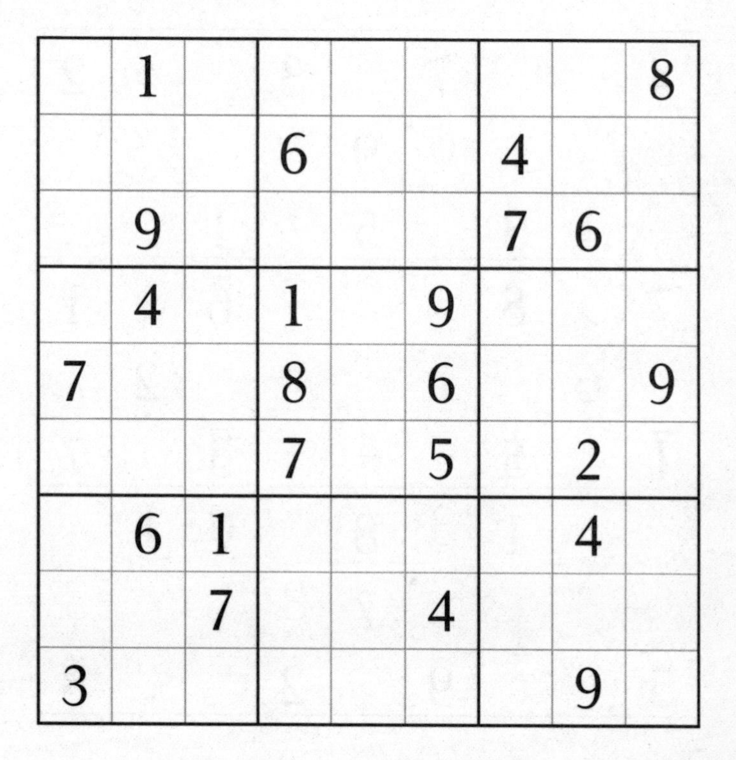

	1							8
			6			4		
	9					7	6	
	4		1		9			
7			8		6			9
			7		5		2	
	6	1					4	
		7			4			
3							9	

Fiendish

98

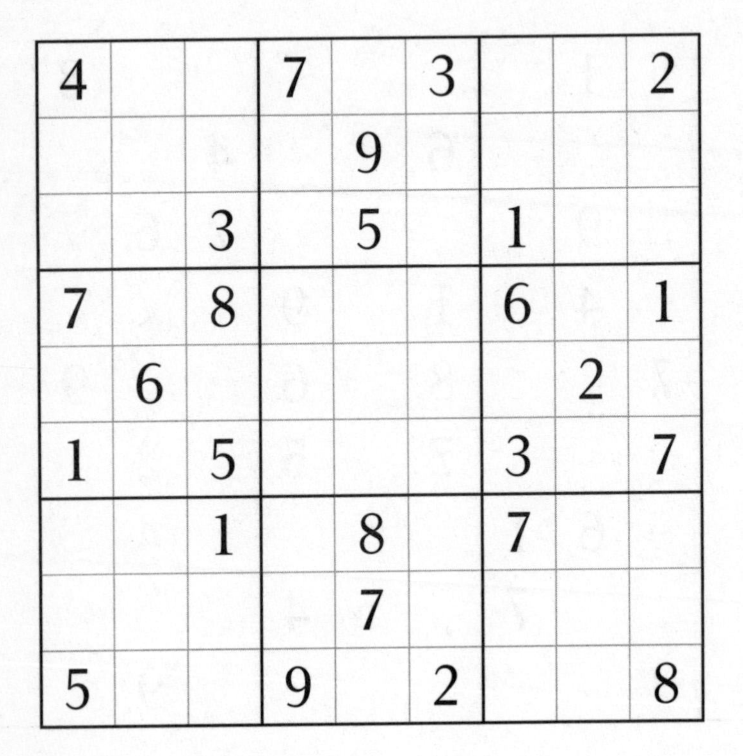

4			7		3			2
				9				
		3		5		1		
7		8				6		1
	6						2	
1		5				3		7
		1		8		7		
				7				
5			9		2			8

Fiendish

99

5								3
		7		1		8		
8		9				2		7
			5		9			
4		3				6		2
			3		6			
9		6				3		8
		8		5		9		
7								1

Fiendish

100

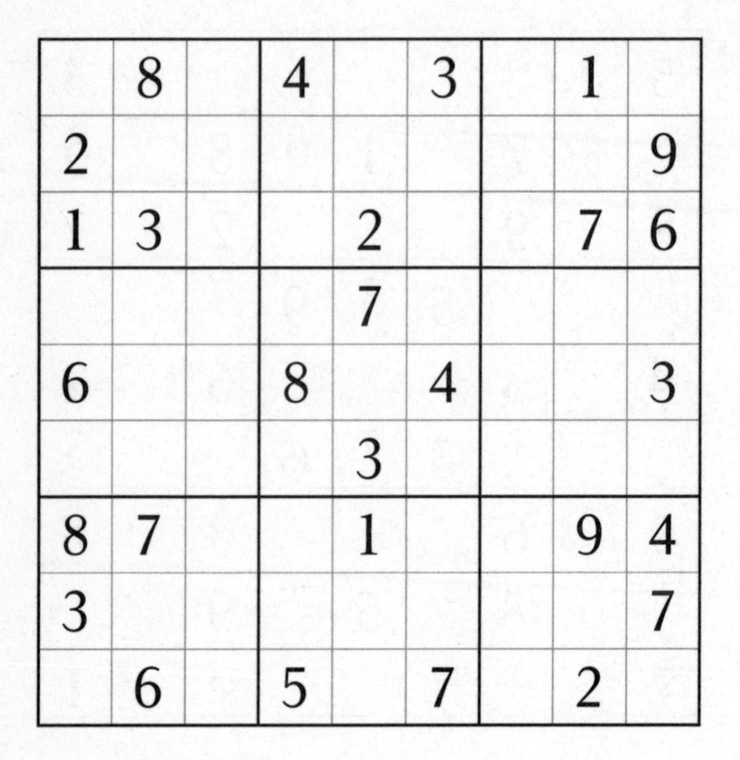

Fiendish

		9			8	5		3
	4				6			8
							4	
6				3	4	9		
		1	2	7				5
	3							
1			5				2	
7		4	9			6		

Fiendish

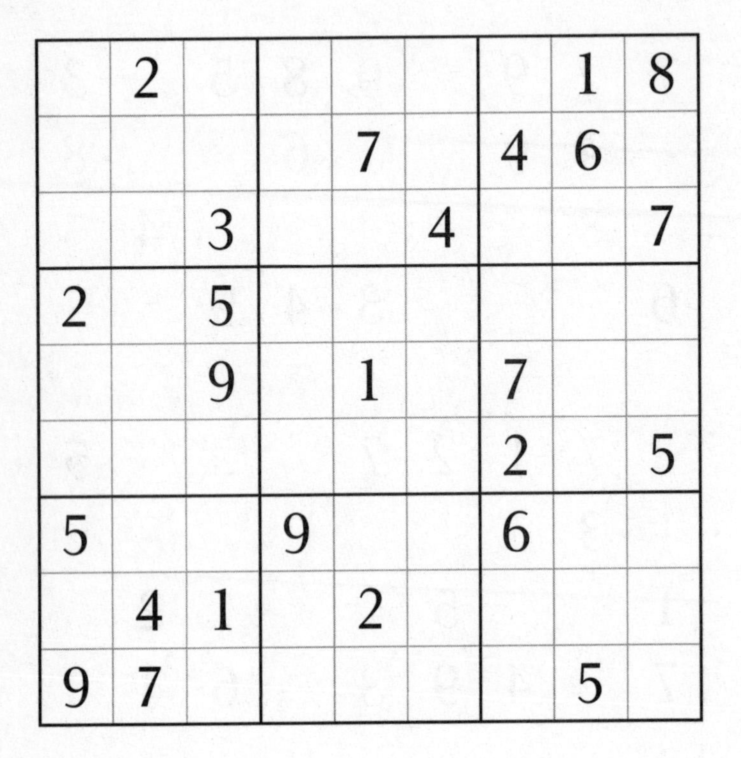

	2						1	8
				7		4	6	
		3			4			7
2		5						
		9		1		7		
						2		5
5			9			6		
	4	1		2				
9	7						5	

Fiendish

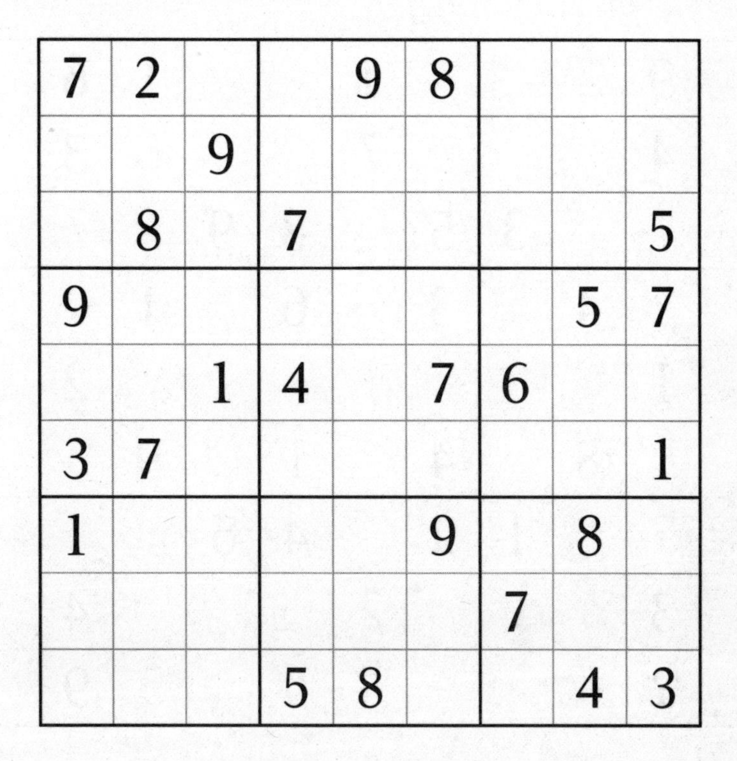

7	2			9	8			
		9						
	8		7					5
9							5	7
		1	4		7	6		
3	7							1
1					9		8	
						7		
			5	8			4	3

Fiendish

104

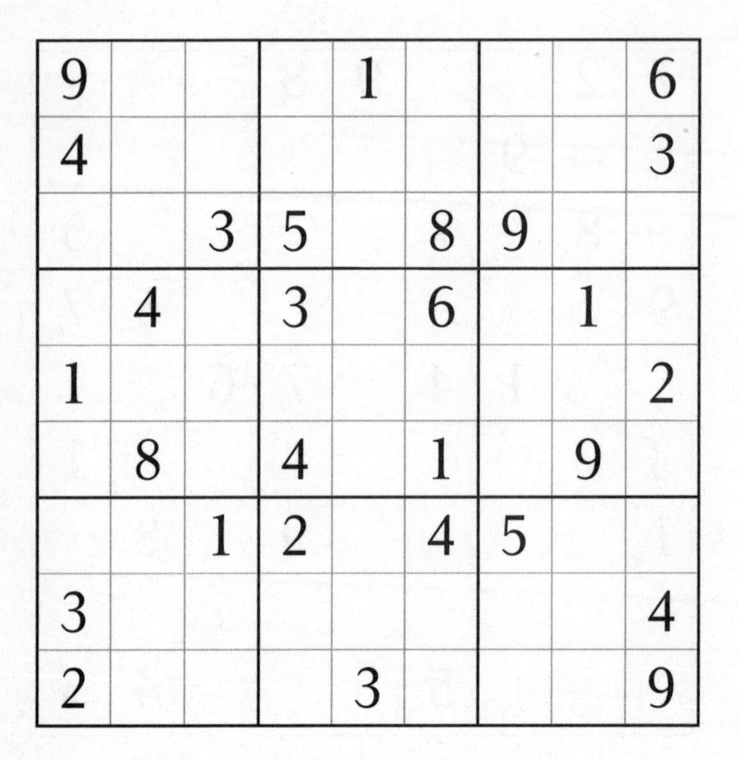

9				1				6
4								3
		3	5		8	9		
	4		3		6		1	
1								2
	8		4		1		9	
		1	2		4	5		
3								4
2				3				9

Fiendish

					4		6	
	9	4	3					
8		7	1					2
		8						1
		5	4		9	8		
6						2		
9					1	5		7
					2	4	1	
	8		7					

Fiendish

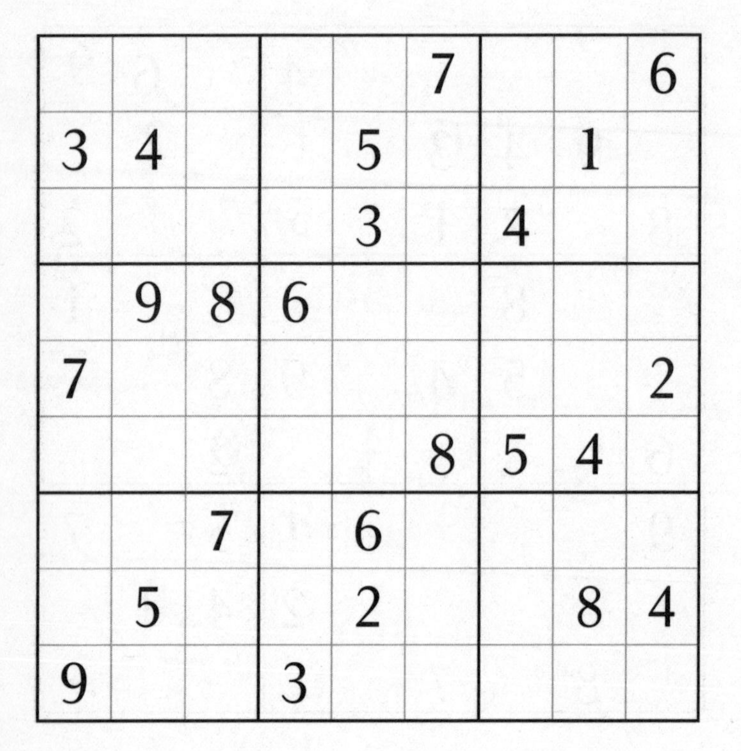

					7			6
3	4			5			1	
				3		4		
	9	8	6					
7								2
					8	5	4	
		7		6				
	5			2			8	4
9			3					

Fiendish

107

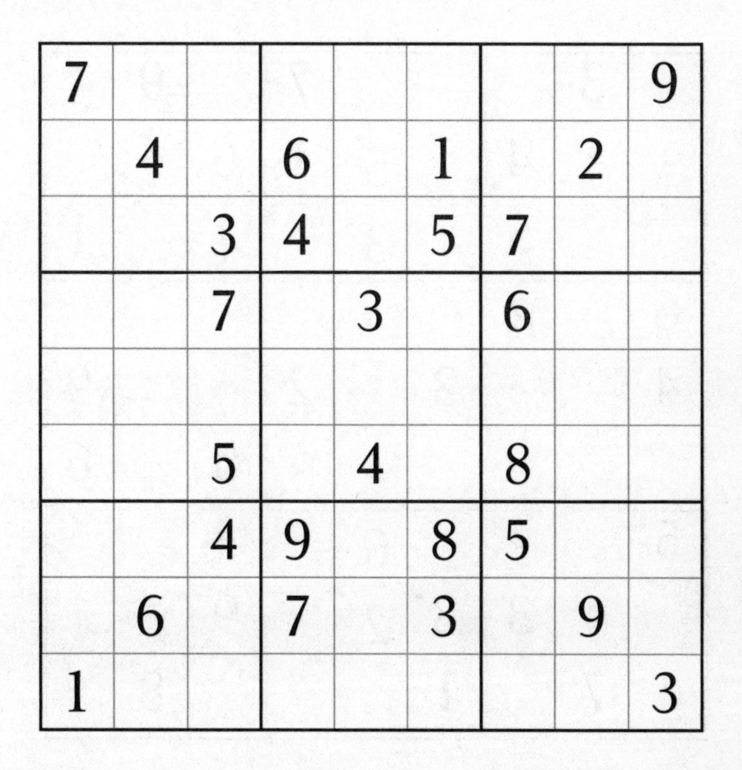

7								9
	4		6		1		2	
		3	4		5	7		
		7		3		6		
		5		4		8		
		4	9		8	5		
	6		7		3		9	
1								3

Fiendish

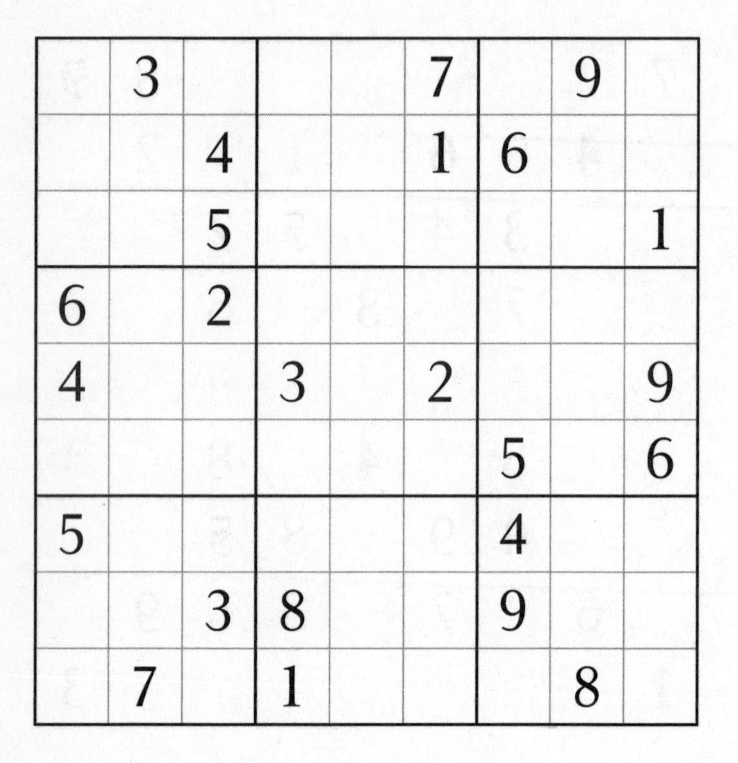

	3				7		9	
		4			1	6		
		5						1
6		2						
4			3		2			9
						5		6
5						4		
		3	8			9		
	7		1				8	

Fiendish

109

	2						8	7
	8	5	6					
		3			7	2		
9				8				3
		6				9		
3				1				4
		9	2			7		
					1	6	2	
2	7						3	

Fiendish

110

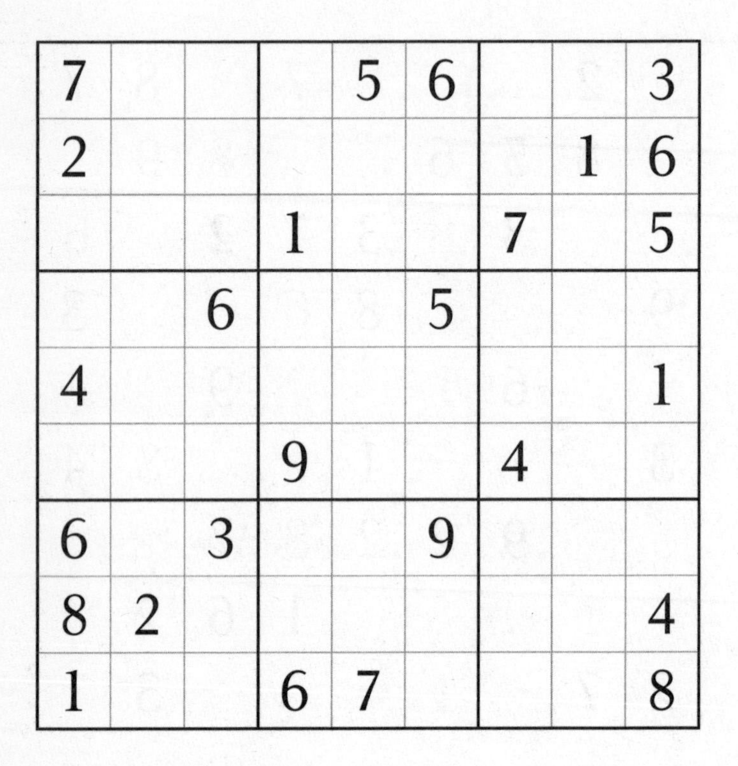

7				5	6			3
2							1	6
			1			7		5
		6			5			
4								1
			9			4		
6		3			9			
8	2							4
1			6	7				8

Fiendish

	4				7			
		7				8	9	
			8	3		4		6
	3				6			1
			7		2			
1			5				8	
6		4		2	8			
	1	9				3		
			1				5	

Fiendish

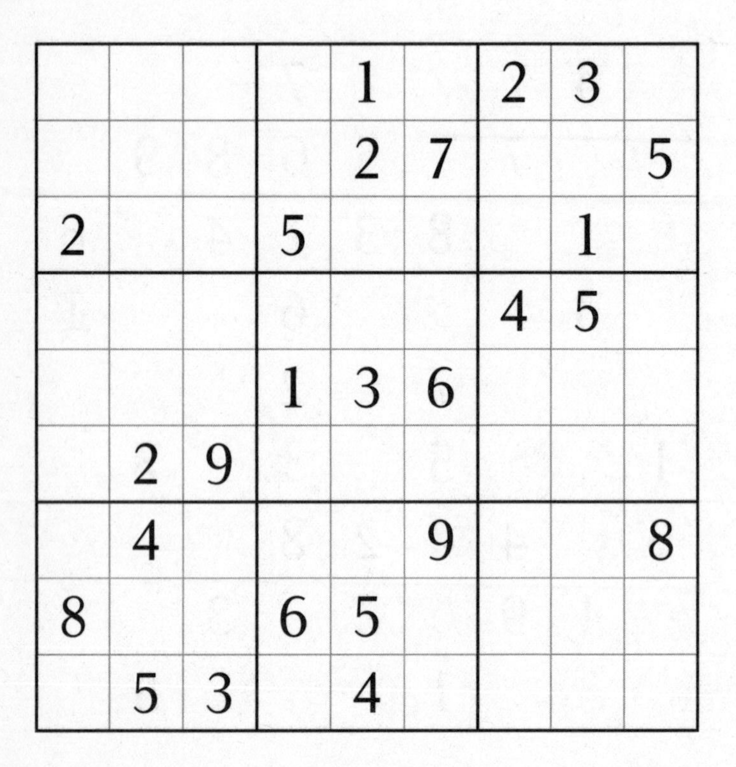

				1		2	3	
				2	7			5
2			5				1	
						4	5	
			1	3	6			
	2	9						
	4				9			8
8			6	5				
	5	3		4				

Fiendish

	6	7						
				5	6		3	
9			2		1			
	3		5		4			8
		9				1		
1			9		3		7	
			4		2			6
	4		6	7				
						2	4	

Fiendish

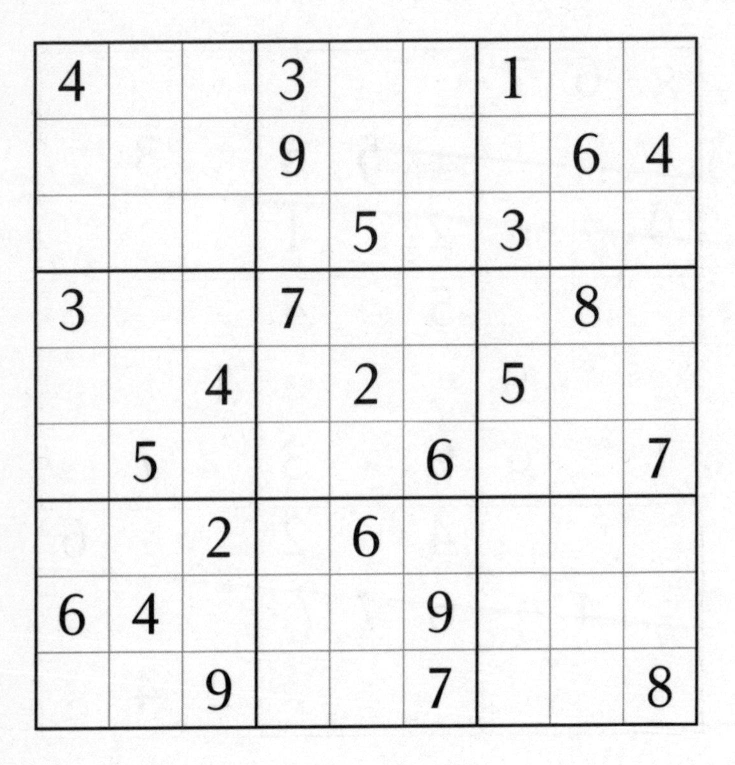

4			3			1		
			9				6	4
				5		3		
3			7				8	
		4		2		5		
	5				6			7
		2		6				
6	4				9			
		9			7			8

Fiendish

115

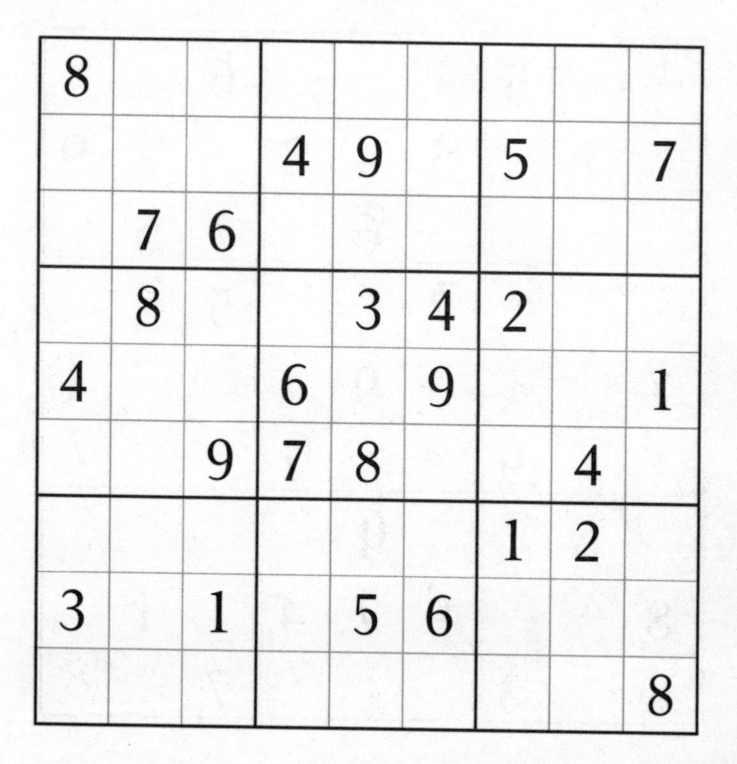

8								
			4	9		5		7
	7	6						
	8			3	4	2		
4			6		9			1
		9	7	8			4	
						1	2	
3		1		5	6			
								8

Fiendish

		5				6		
	7		5	4	3			9
				6			5	
			7			5		
7				9				1
		9			2			
	6			1				
8			9	7	4		1	
		3				7		

Fiendish

5		3	6					9
	1				2	6		
9							8	
			7					5
		6	8		4	1		
2					3			
	3							8
		4	3				5	
8					6	7		2

Fiendish

118

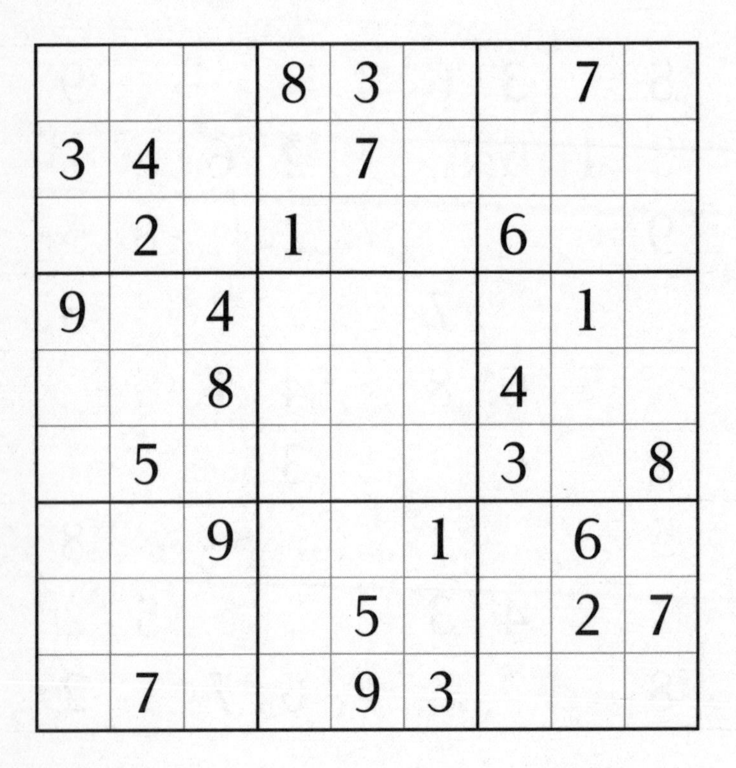

			8	3			7	
3	4			7				
	2		1			6		
9		4					1	
		8				4		
	5					3		8
		9			1		6	
				5			2	7
	7			9	3			

Fiendish

8		5		3				
3		6			4			1
				7	2			5
			4			2		
	5						8	
		4			1			
5			2	6				
7			3			5		2
				1		4		9

Fiendish

120

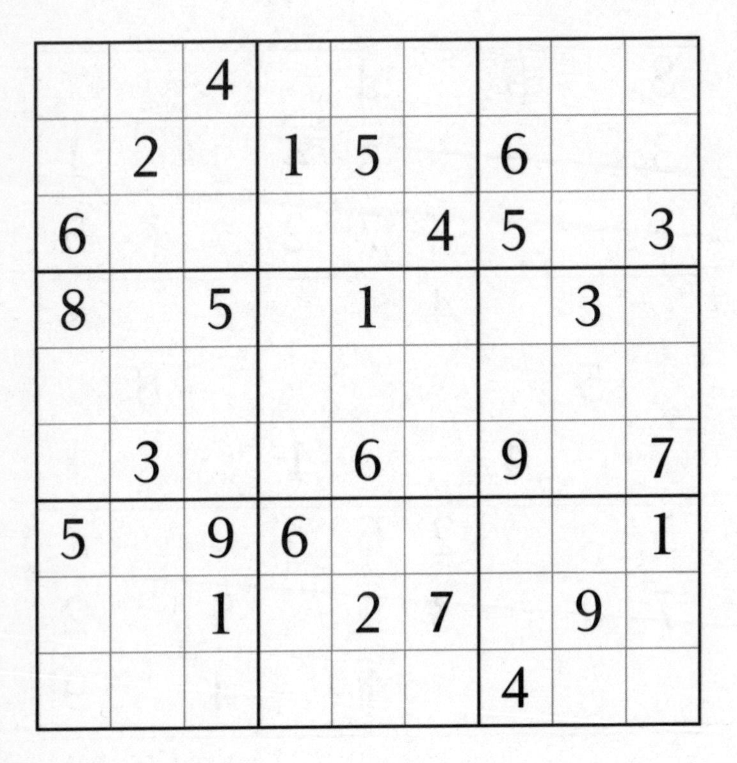

		4						
	2		1	5		6		
6					4	5		3
8		5		1			3	
	3			6		9		7
5		9	6					1
		1		2	7		9	
						4		

Fiendish

121

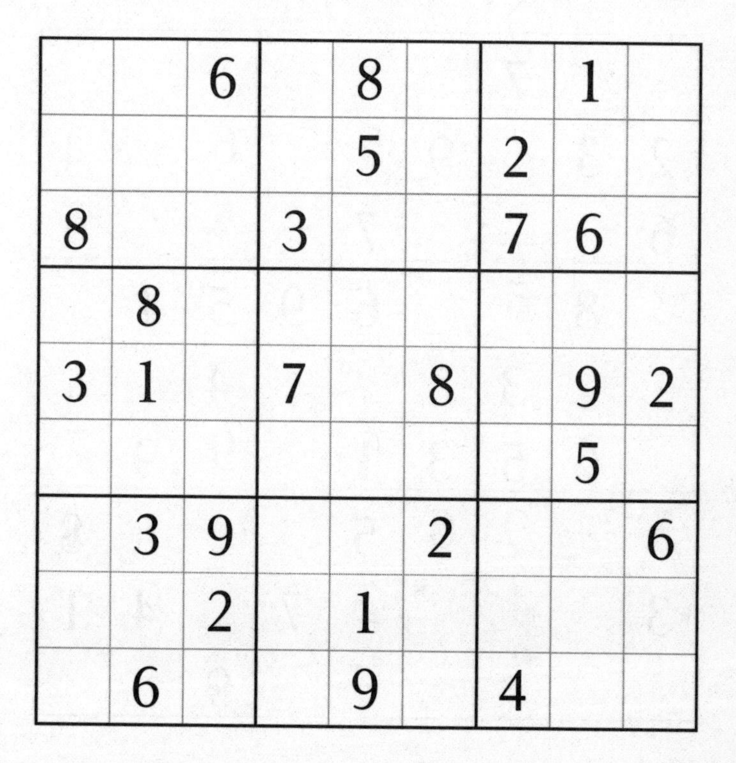

		6		8			1	
				5		2		
8			3			7	6	
	8							
3	1		7		8		9	2
							5	
	3	9			2			6
		2		1				
	6			9		4		

Fiendish

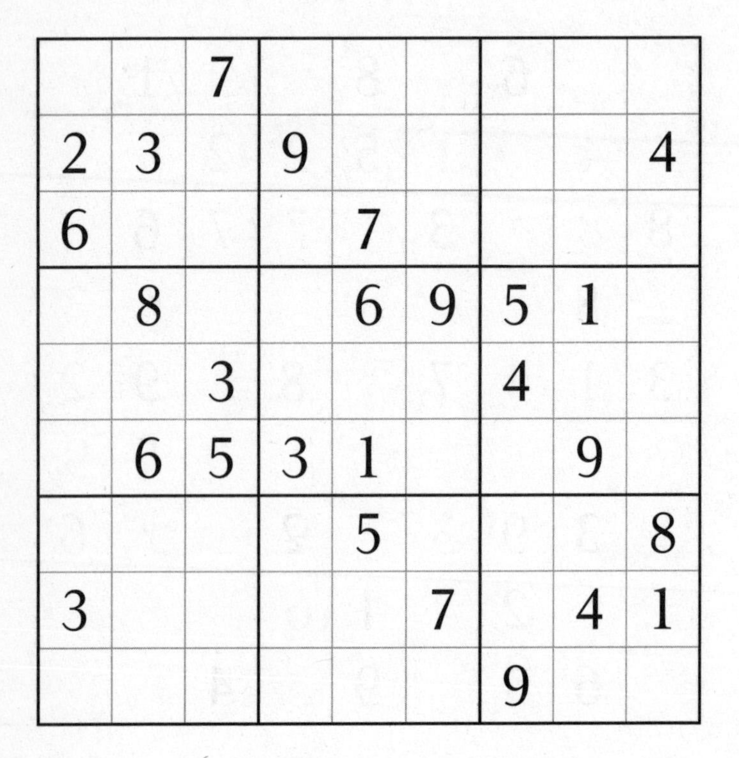

		7						
2	3		9					4
6				7				
	8			6	9	5	1	
		3				4		
	6	5	3	1			9	
				5				8
3					7		4	1
						9		

Fiendish

		2		8		9		
	4		1		2		5	
	8		5		7		3	
2								3
		4				8		
7								9
	2		3		9		7	
	1		4		6		8	
		5		2		3		

Fiendish

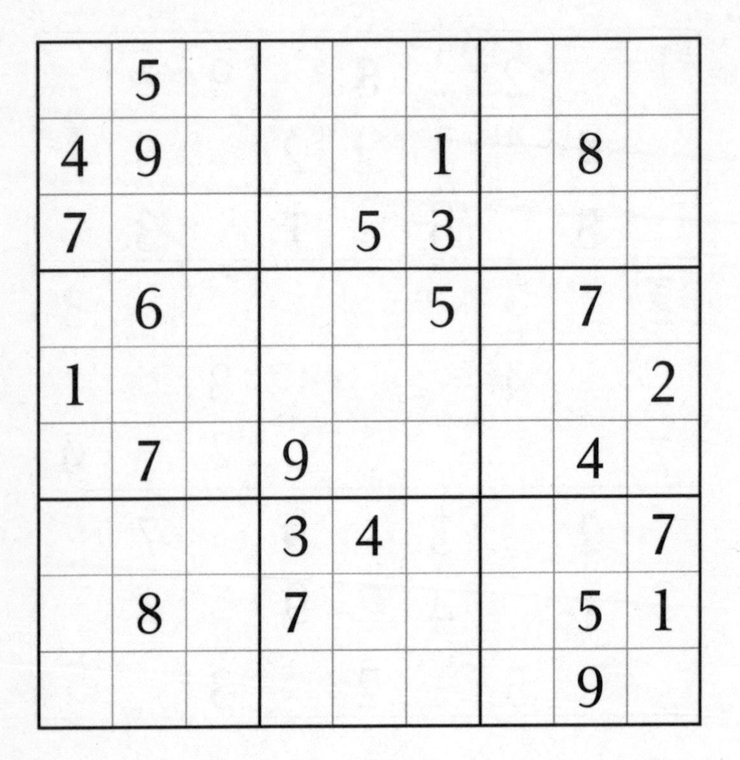

	5							
4	9				1		8	
7				5	3			
	6				5		7	
1								2
	7		9				4	
			3	4				7
	8		7				5	1
							9	

125

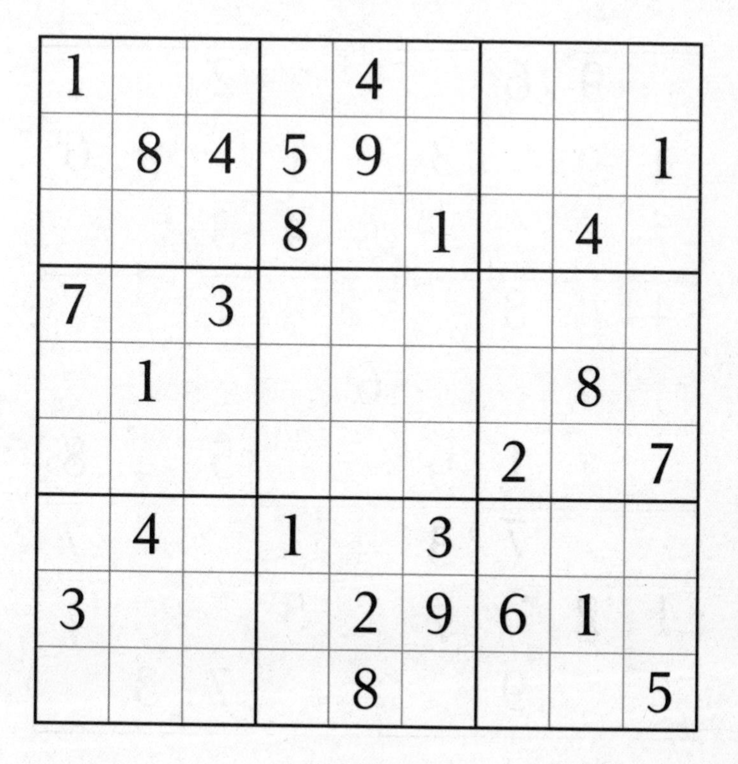

1				4				
	8	4	5	9				1
			8		1		4	
7		3						
	1						8	
						2		7
	4		1		3			
3				2	9	6	1	
				8				5

Fiendish

126

	9	6				2		
			3				9	6
					4	3		
4		8				1		
	5			6			7	
		1				5		8
		7	4					
1	2				5			
		9				7	3	

Fiendish

		8		9				
					1		4	
4	2		3			8		
		6						9
7			6		2			3
1						5		
		9			6		8	4
	5		4					
				7		2		

Fiendish

128

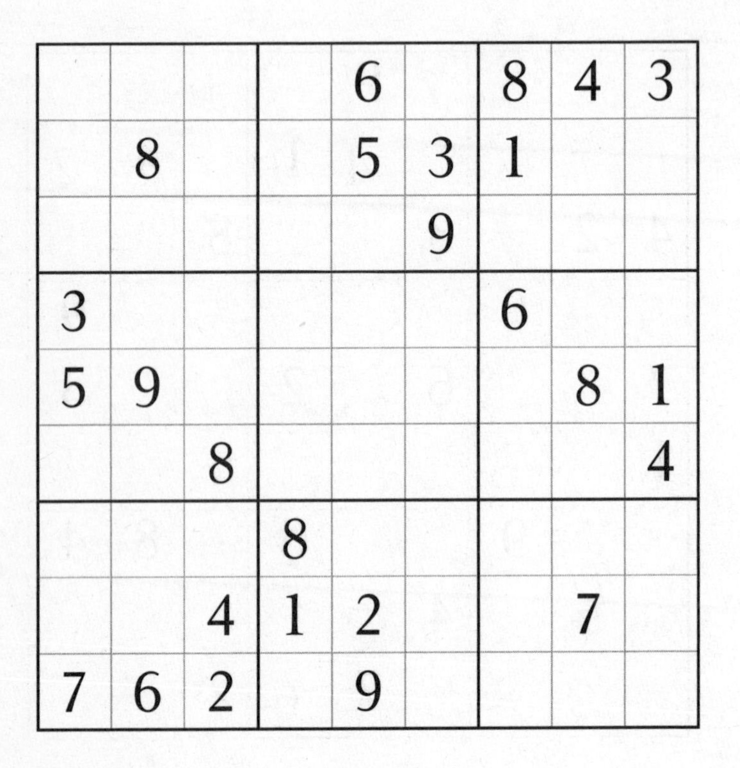

				6		8	4	3
	8			5	3	1		
					9			
3						6		
5	9						8	1
		8						4
			8					
		4	1	2			7	
7	6	2		9				

Fiendish

129

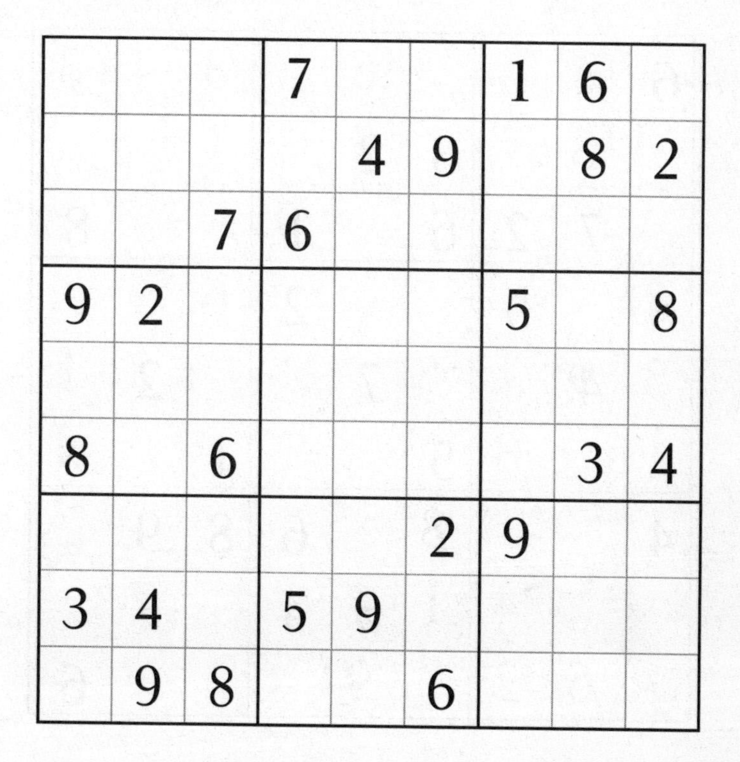

			7			1	6	
				4	9		8	2
		7	6					
9	2					5		8
8		6					3	4
					2	9		
3	4		5	9				
	9	8			6			

Fiendish

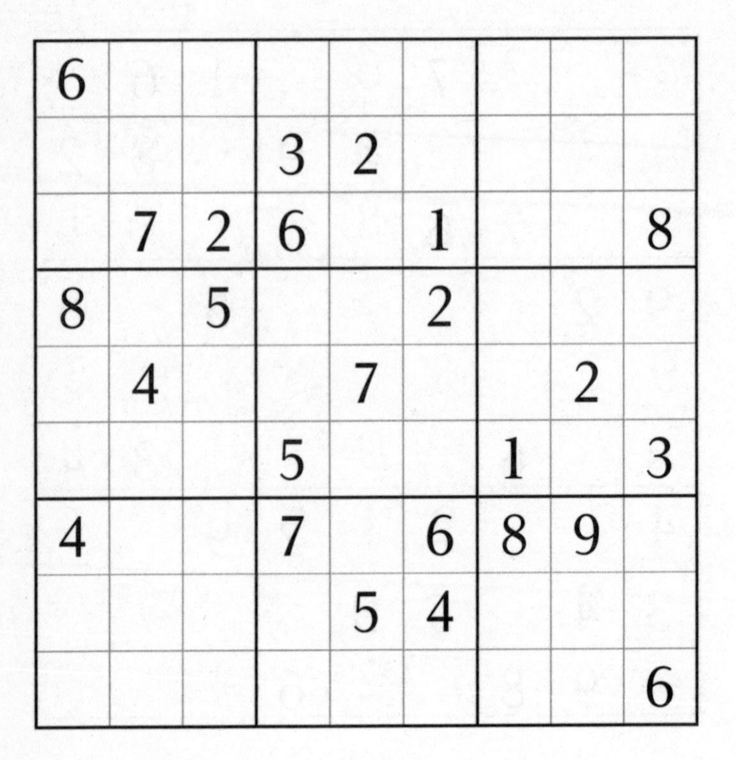

Fiendish

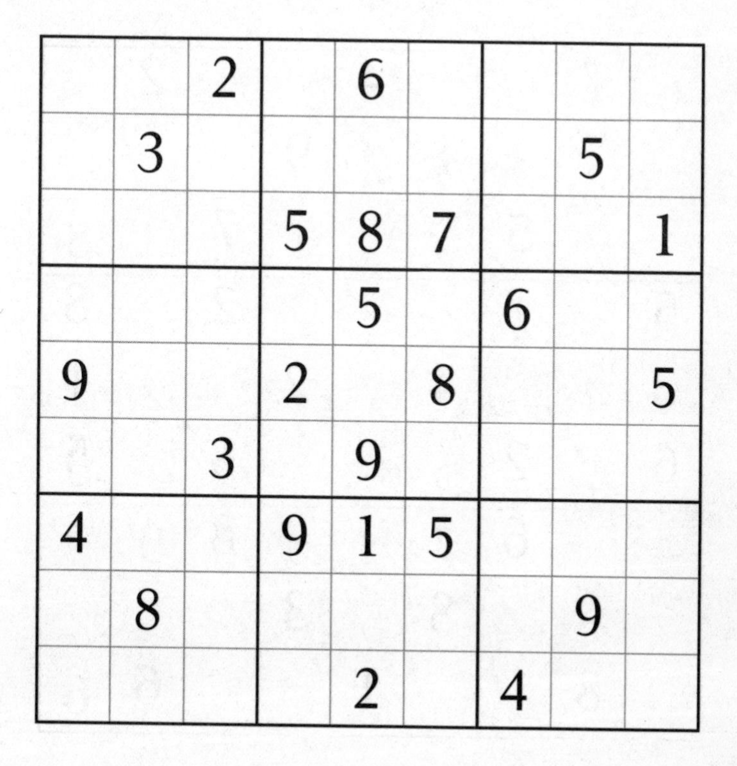

		2		6				
	3						5	
			5	8	7			1
				5		6		
9			2		8			5
		3		9				
4			9	1	5			
	8						9	
				2		4		

Fiendish

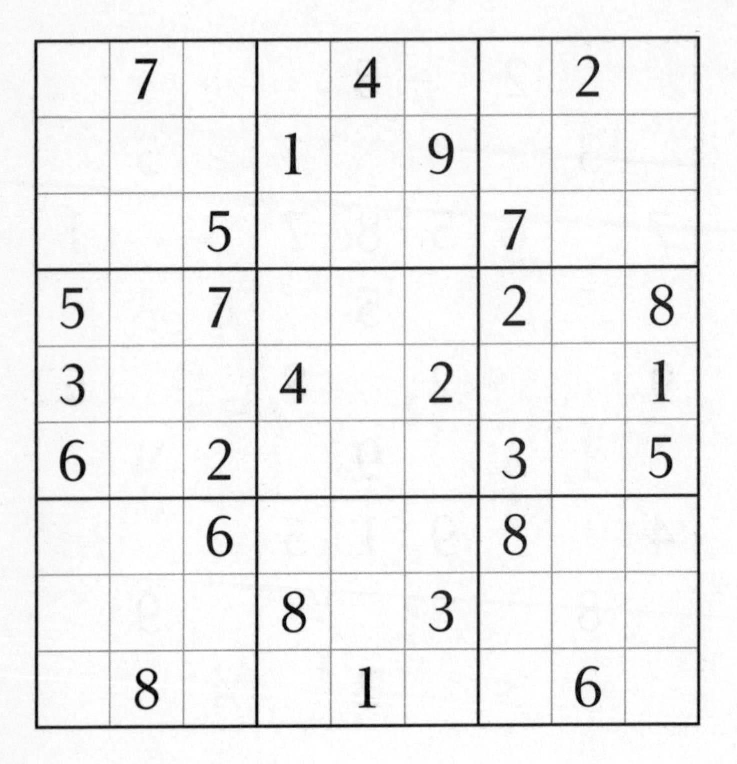

	7			4			2	
			1		9			
		5				7		
5		7				2		8
3			4		2			1
6		2				3		5
		6				8		
			8		3			
	8			1			6	

133

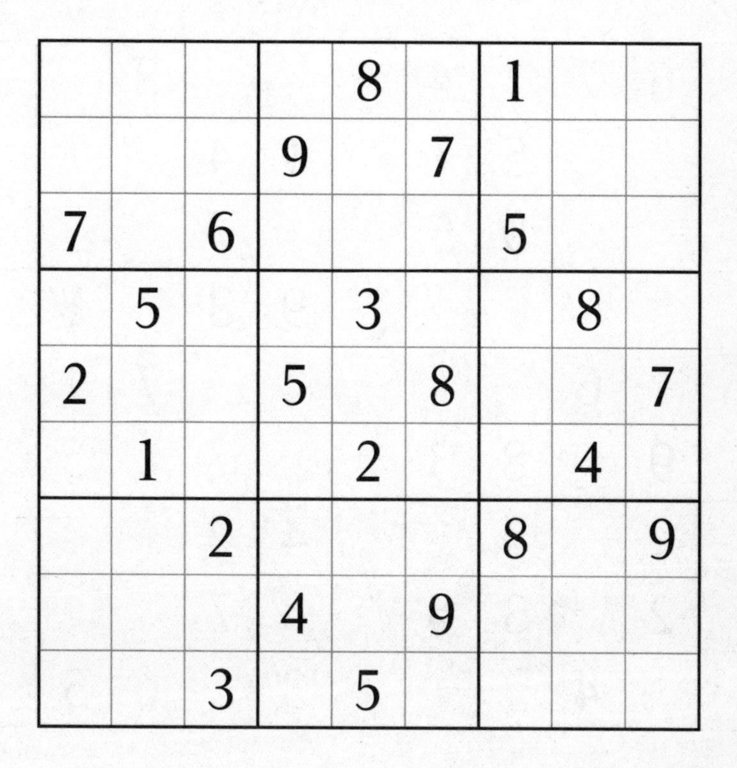

				8		1		
			9		7			
7		6				5		
	5			3			8	
2			5		8			7
	1			2			4	
		2				8		9
			4		9			
		3		5				

Fiendish

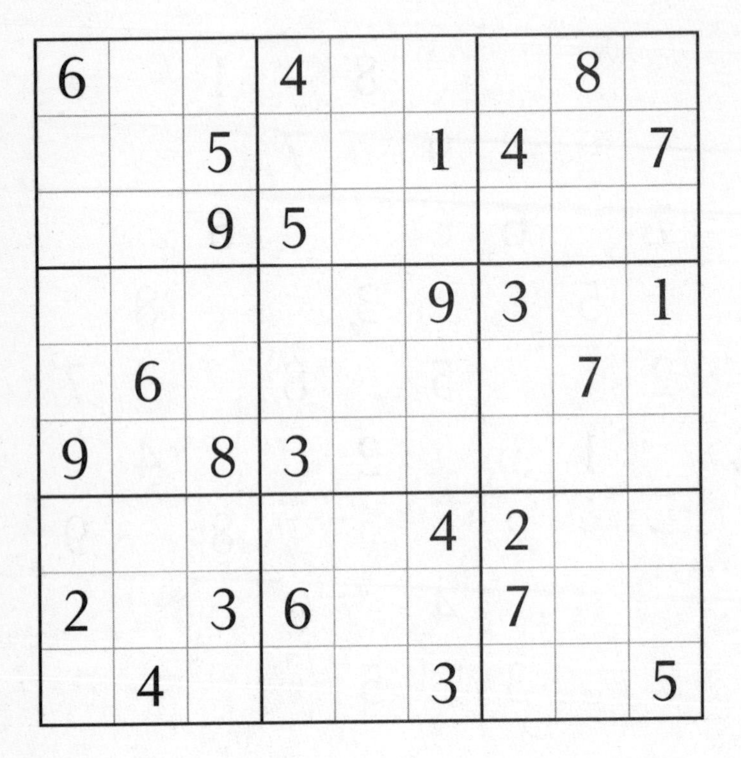

6			4				8	
		5			1	4		7
		9	5					
					9	3		1
	6						7	
9		8	3					
					4	2		
2		3	6			7		
	4				3			5

Fiendish

7						9		6
			7	3				
8			9		5			
		5				8	4	
	1						9	
	3	7				1		
			4		3			8
				6	2			
5		1						3

Fiendish

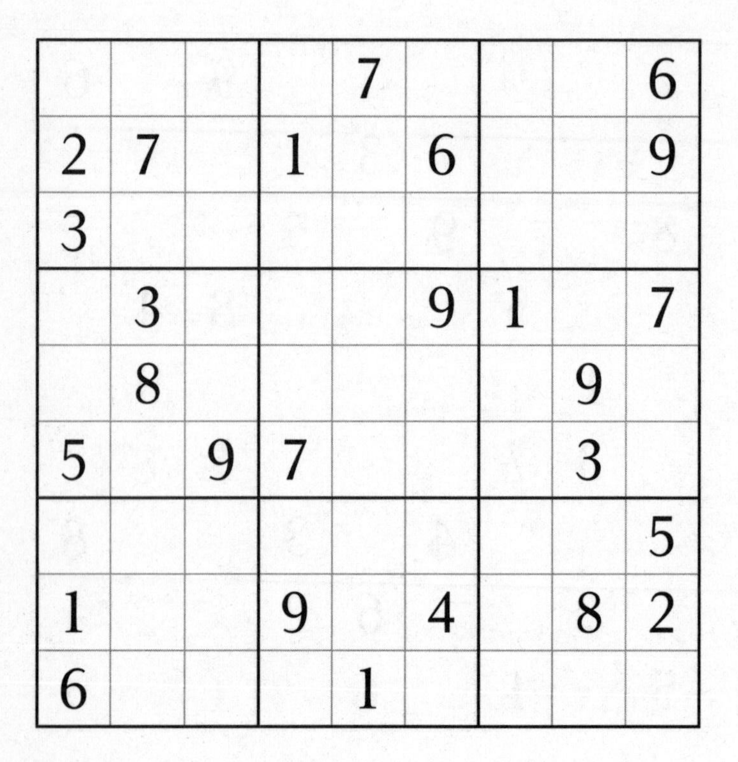

				7				6
2	7		1		6			9
3								
	3				9	1		7
	8						9	
5		9	7				3	
								5
1			9		4		8	2
6				1				

Fiendish

137

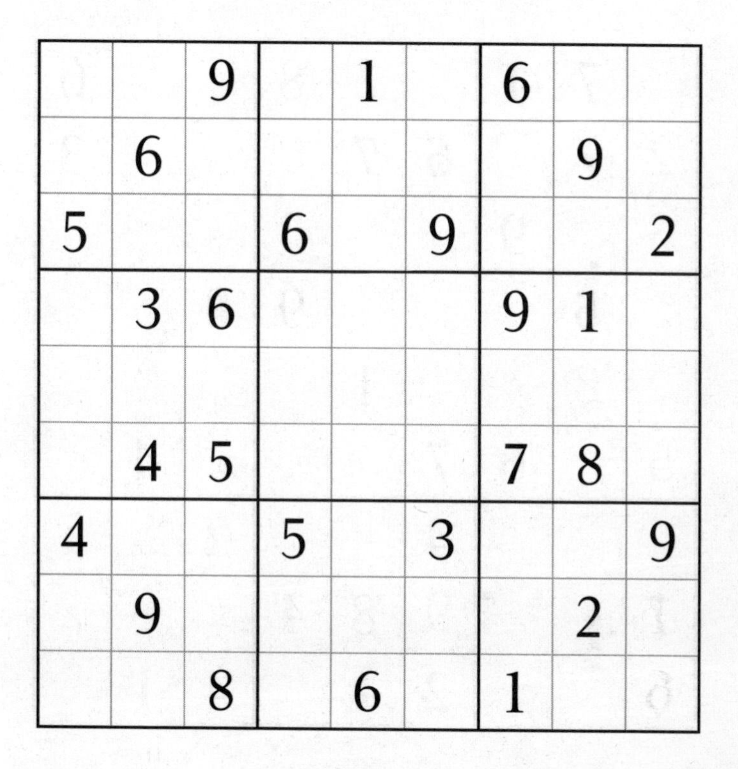

		9		1		6		
	6						9	
5			6		9			2
	3	6				9	1	
	4	5				7	8	
4			5		3			9
	9						2	
		8		6		1		

Fiendish

138

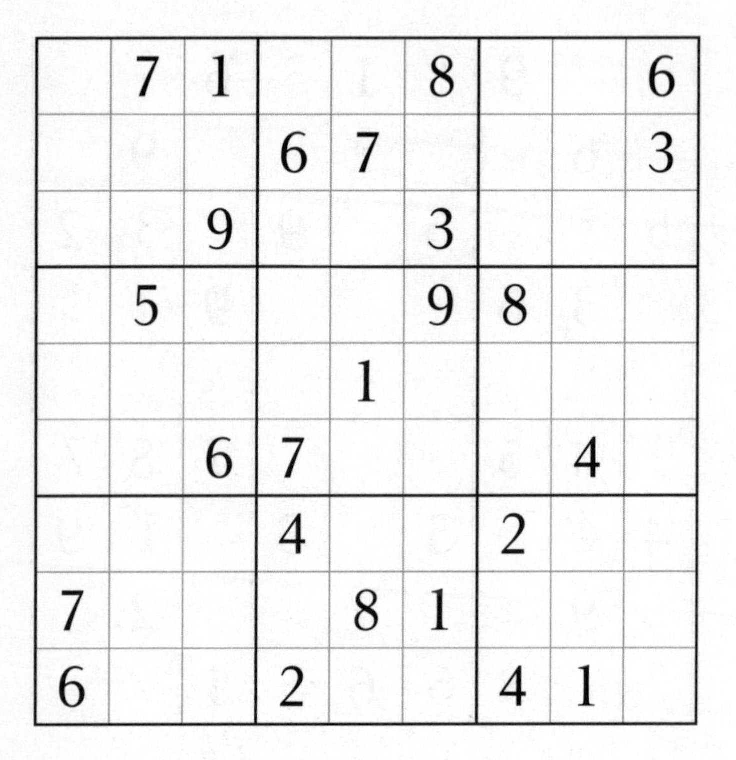

	7	1			8			6
			6	7				3
		9			3			
	5				9	8		
				1				
		6	7				4	
			4			2		
7				8	1			
6			2			4	1	

Fiendish

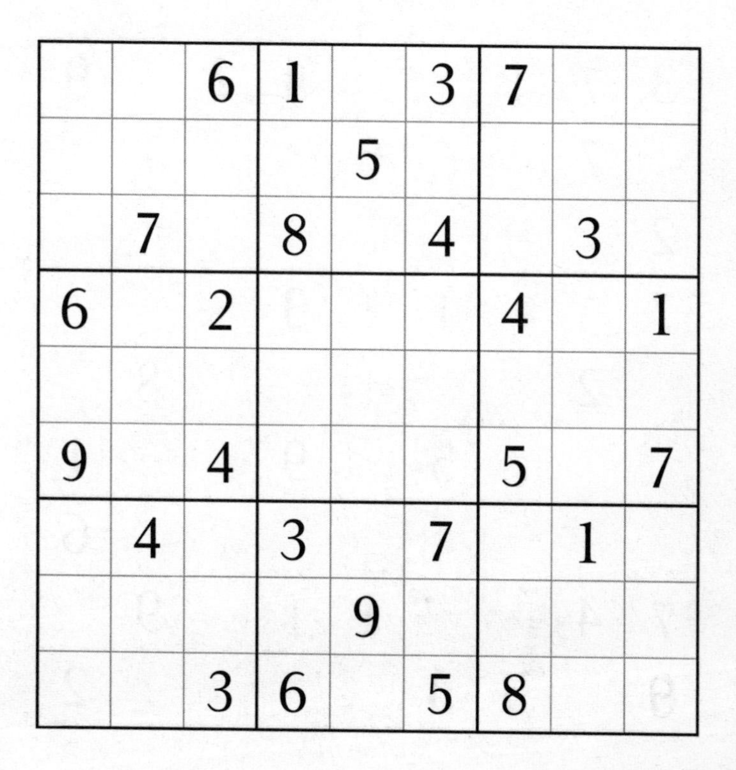

		6	1		3	7		
				5				
	7		8		4		3	
6		2				4		1
9		4				5		7
	4		3		7		1	
				9				
		3	6		5	8		

Fiendish

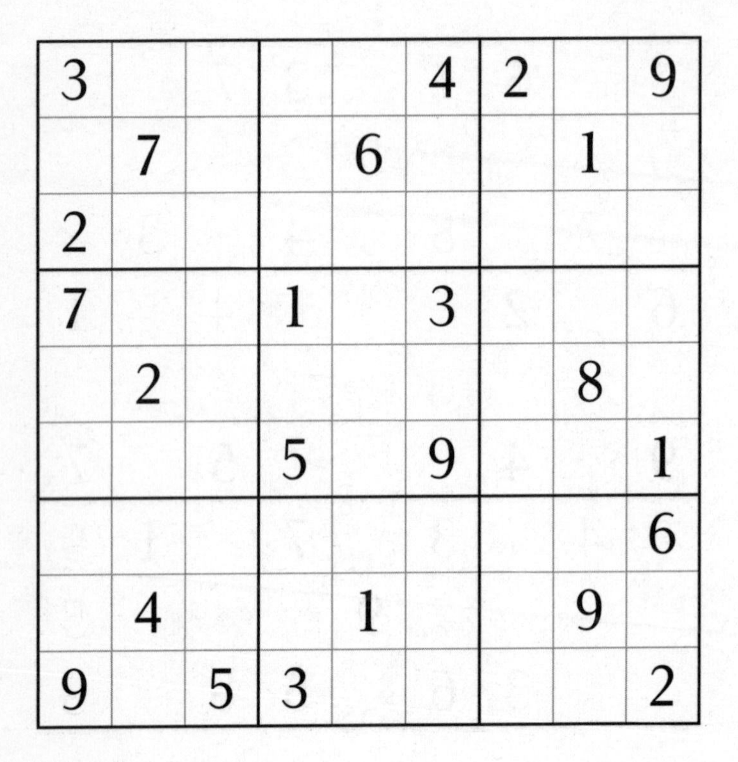

3					4	2		9
	7			6			1	
2								
7			1		3			
	2						8	
			5		9			1
								6
	4			1			9	
9		5	3					2

Fiendish

141

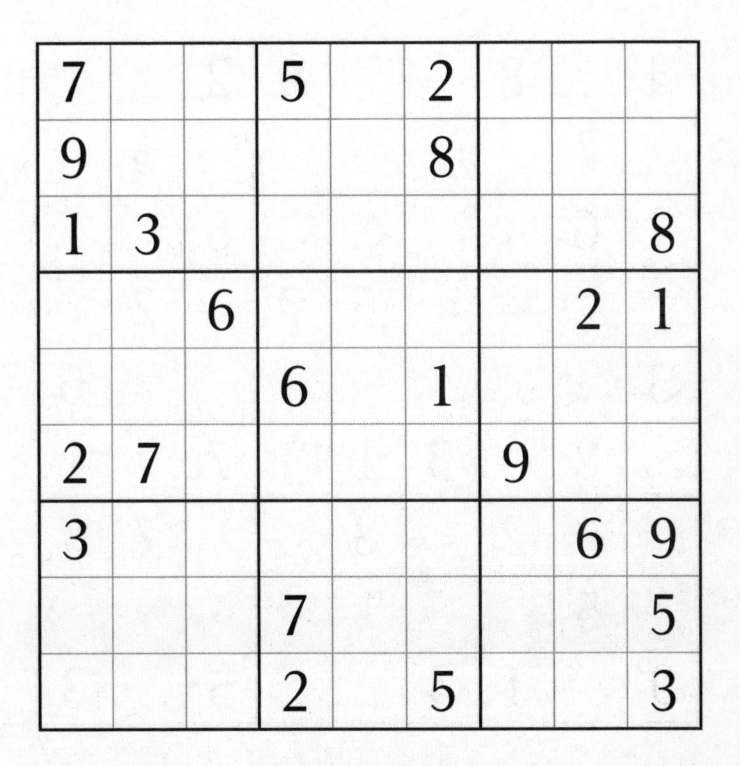

7			5		2			
9					8			
1	3							8
		6					2	1
			6		1			
2	7					9		
3							6	9
			7					5
			2		5			3

Fiendish

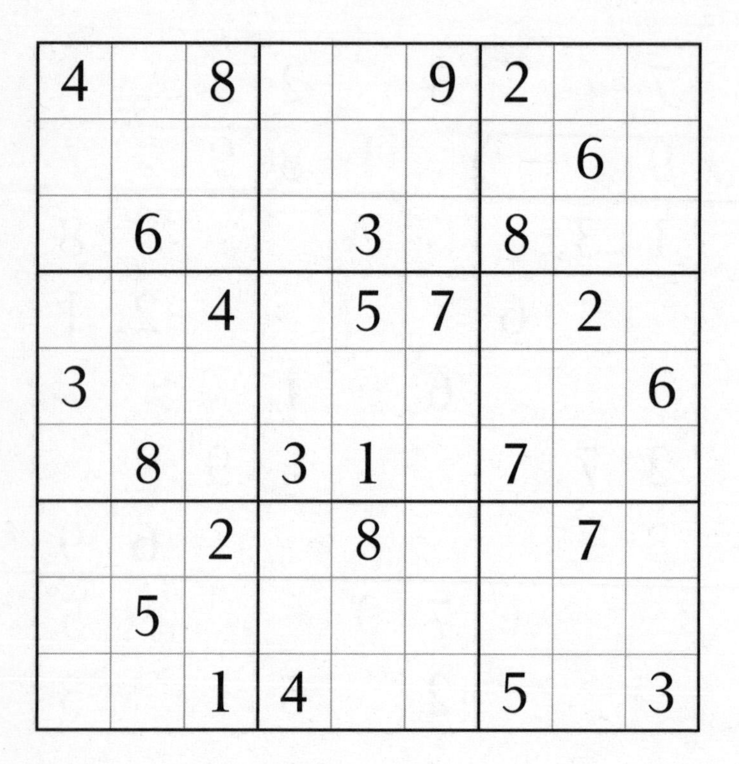

4		8			9	2		
							6	
	6			3		8		
		4		5	7		2	
3								6
	8		3	1		7		
		2		8			7	
	5							
		1	4			5		3

Fiendish

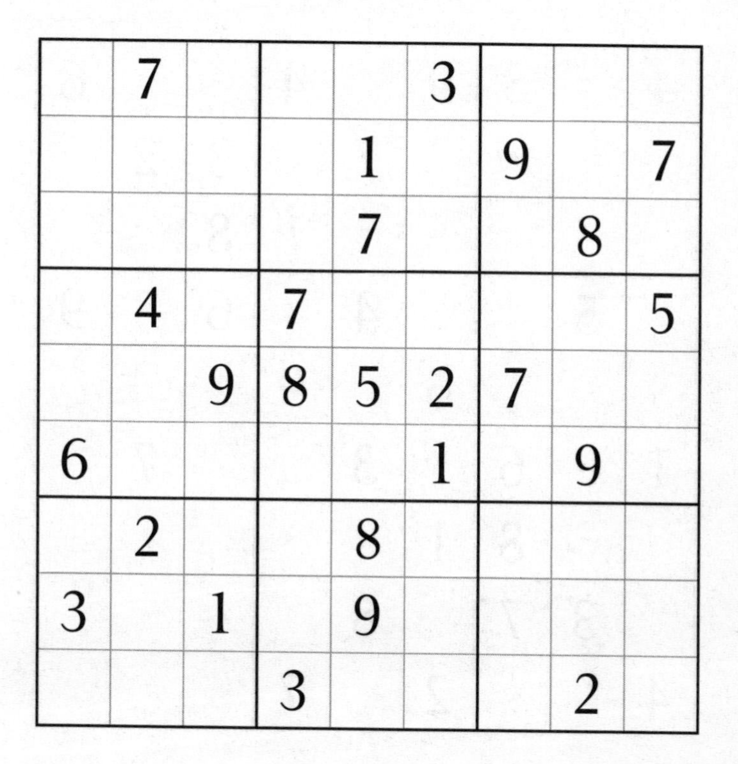

	7				3			
				1		9		7
				7			8	
	4		7					5
		9	8	5	2	7		
6					1		9	
	2			8				
3		1		9				
			3				2	

Fiendish

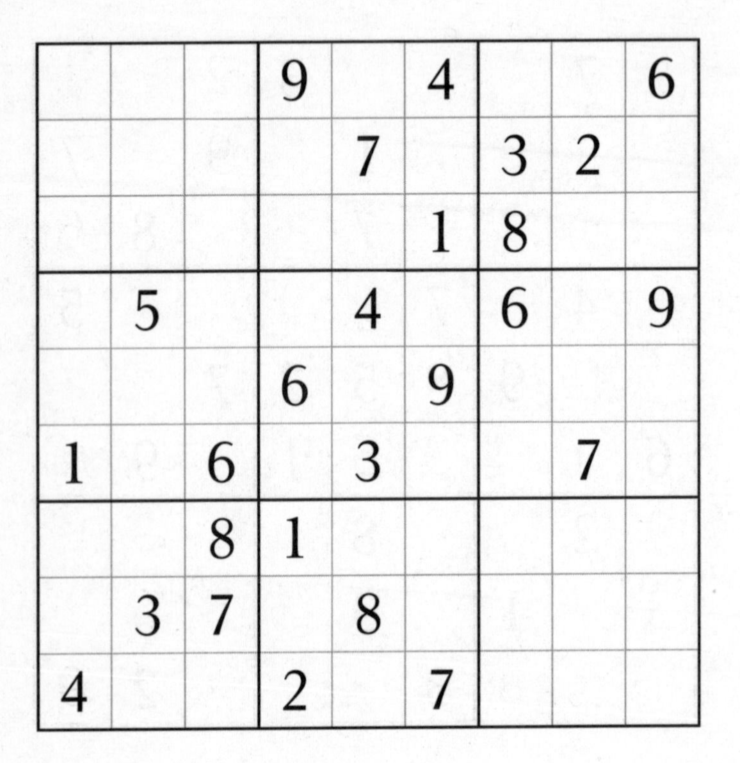

			9		4			6
				7		3	2	
					1	8		
	5			4		6		9
			6		9			
1		6		3			7	
		8	1					
	3	7		8				
4			2		7			

Fiendish

5						2		1
	4	6	7					
	1							6
				7	9			3
		3				4		
8			3	5				
9							2	
					2	8	6	
1		8						9

Fiendish

146

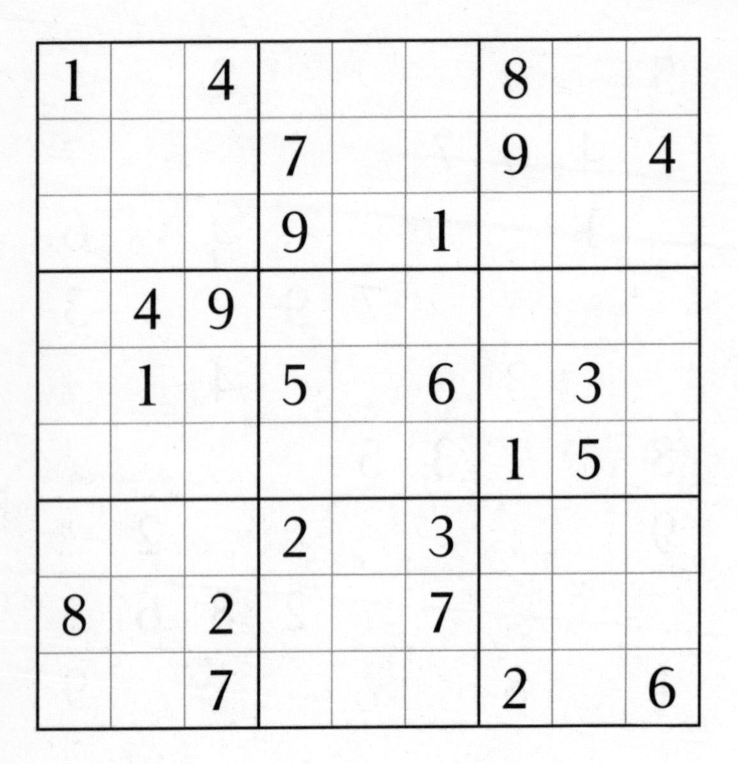

1		4				8		
			7			9		4
			9		1			
	4	9						
	1		5		6		3	
						1	5	
			2		3			
8		2			7			
		7				2		6

Fiendish

147

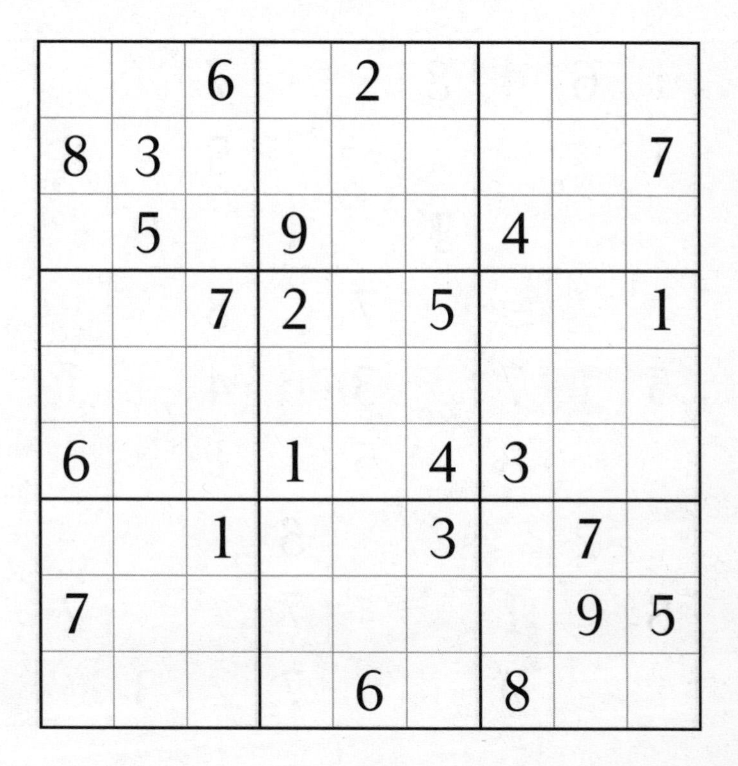

		6		2				
8	3							7
	5		9			4		
		7	2		5			1
6			1		4	3		
		1			3		7	
7							9	5
				6		8		

Fiendish

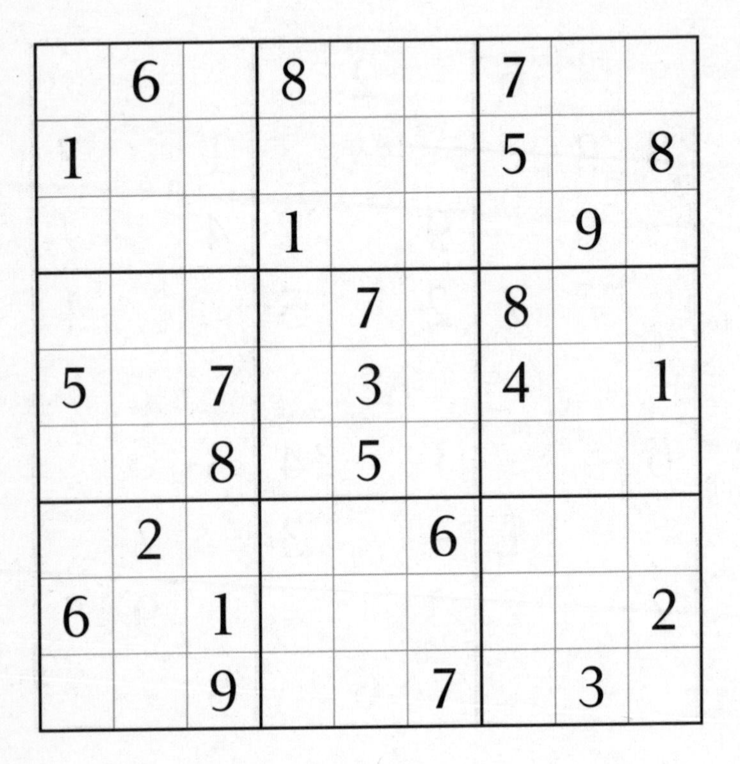

	6		8			7		
1						5		8
			1				9	
				7		8		
5		7		3		4		1
		8		5				
	2				6			
6		1						2
		9			7		3	

Fiendish

	1							
		8			3	1		2
					2	4		7
	2		6			7		
4			2		8			9
		3			5		6	
8		9	4					
1		2	3			5		
							7	

Fiendish

150

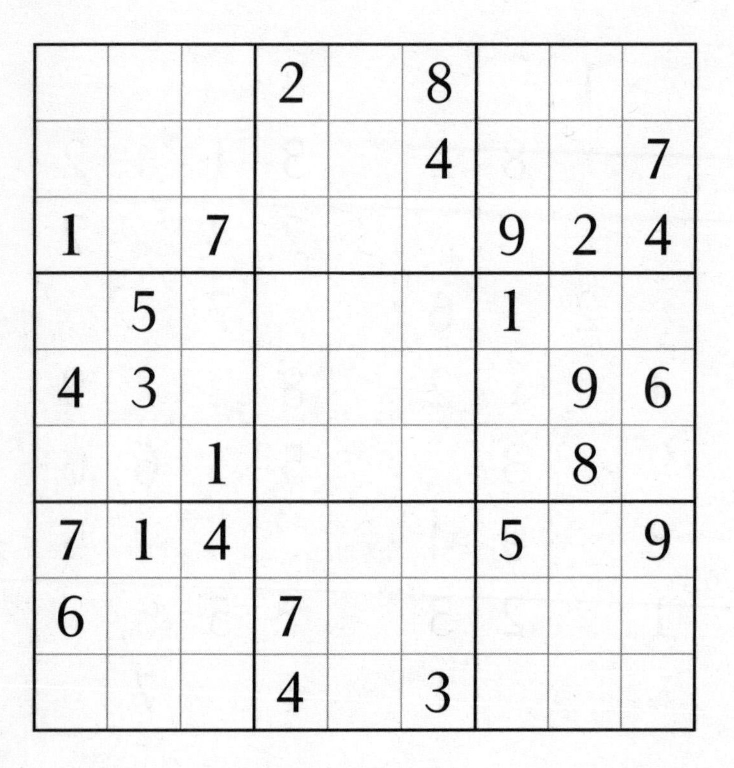

			2		8			
					4			7
1		7				9	2	4
	5					1		
4	3						9	6
		1					8	
7	1	4				5		9
6			7					
			4		3			

Fiendish

151

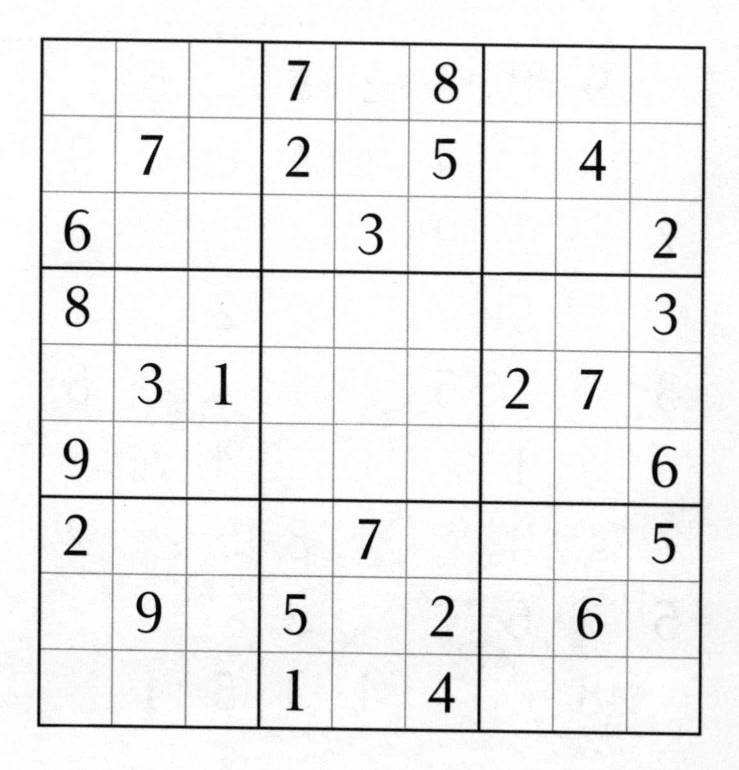

			7		8			
	7		2		5		4	
6				3				2
8								3
	3	1				2	7	
9								6
2				7				5
	9		5		2		6	
			1		4			

Fiendish

	9	8		2			6	
			4			8		9
7			9					
		5				2		
3			5		1			8
		1				4		
					8			2
5		6			3			
	8			4		5	1	

Fiendish

		9		8		3		
	8				5			2
1		4	6			9		
				6			1	
			7		2			
	7			3				
		3			6	4		9
8			5				6	
		6		1		7		

Fiendish

154

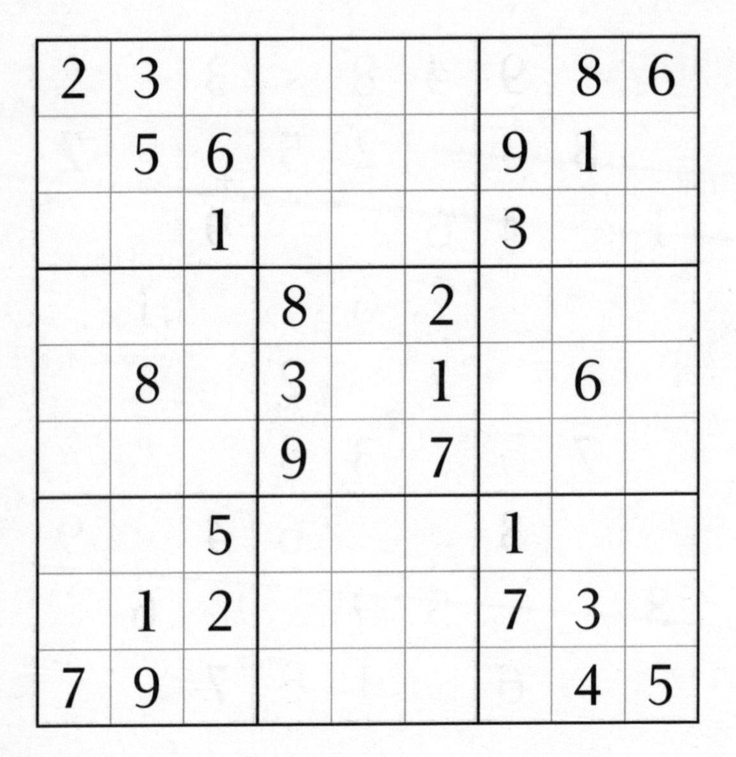

2	3						8	6
	5	6				9	1	
		1				3		
			8		2			
	8		3		1		6	
			9		7			
		5				1		
	1	2				7	3	
7	9						4	5

Fiendish

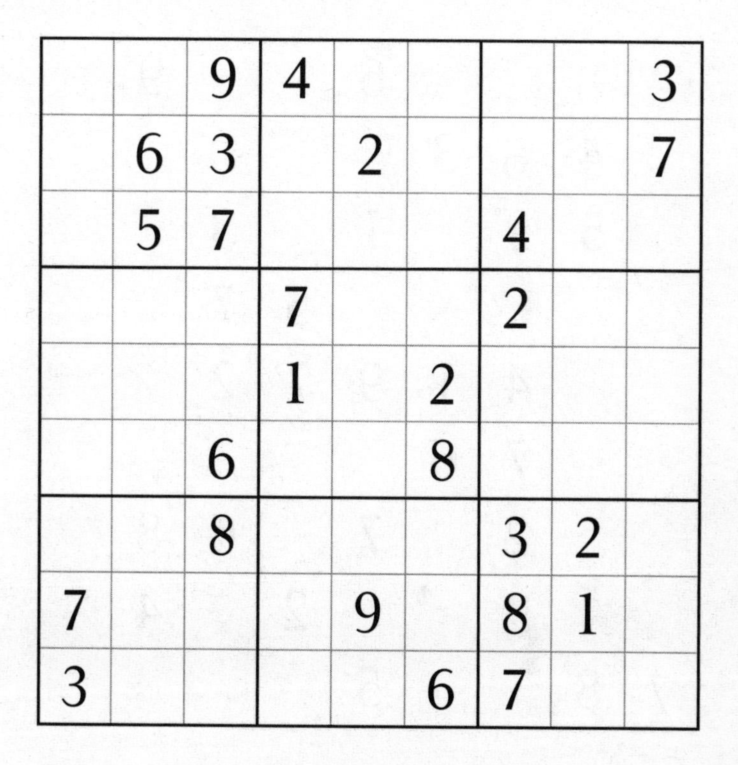

		9	4					3
	6	3		2				7
	5	7				4		
			7			2		
			1		2			
		6			8			
		8				3	2	
7				9		8	1	
3					6	7		

Fiendish

156

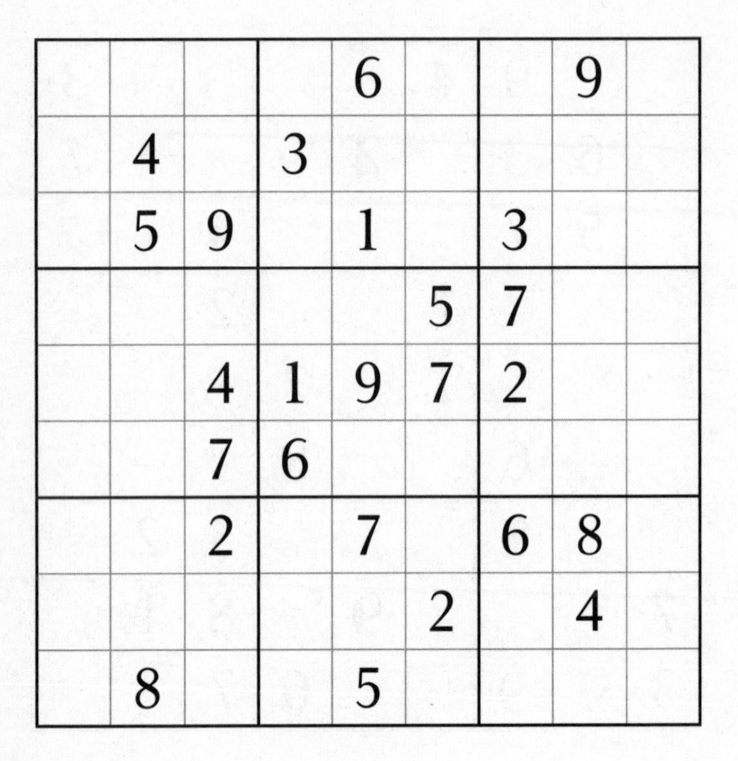

				6			9	
	4		3					
	5	9		1		3		
					5	7		
		4	1	9	7	2		
		7	6					
		2		7		6	8	
					2		4	
	8			5				

Fiendish

157

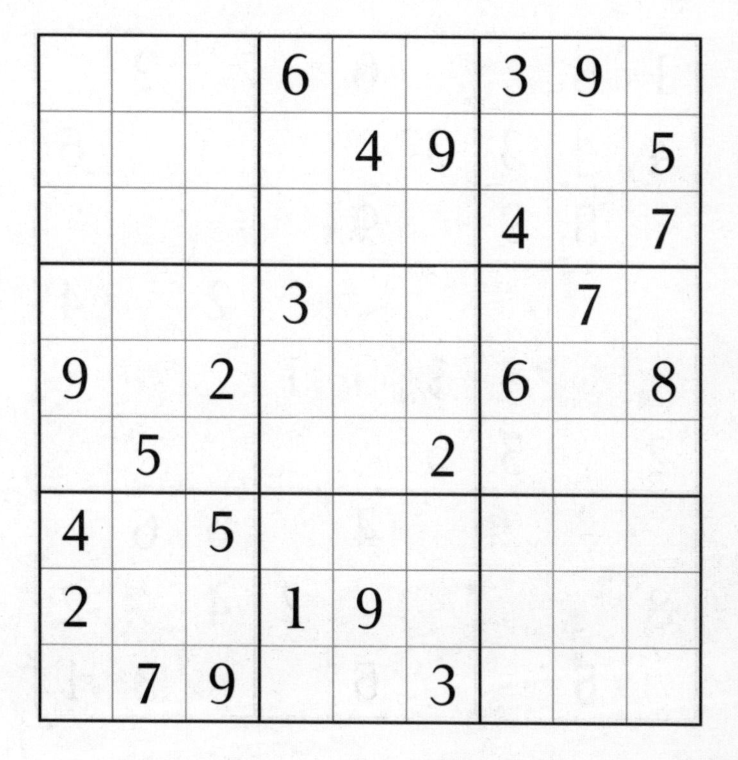

			6			3	9	
				4	9			5
						4		7
			3				7	
9		2				6		8
	5				2			
4		5						
2			1	9				
	7	9			3			

Fiendish

158

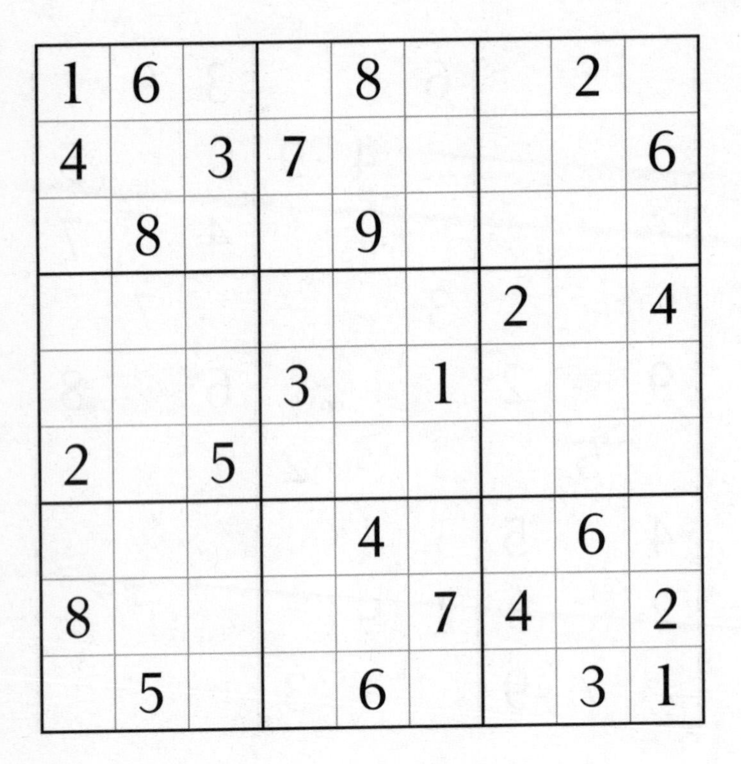

1	6			8			2	
4		3	7					6
	8			9				
						2		4
			3		1			
2		5						
				4			6	
8					7	4		2
	5			6			3	1

Fiendish

159

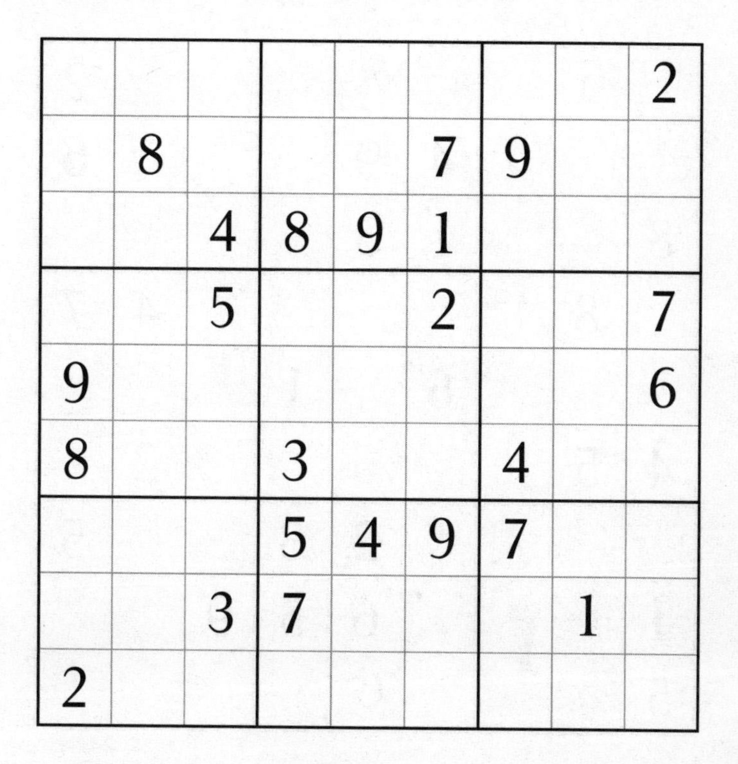

								2
	8				7	9		
		4	8	9	1			
		5			2			7
9								6
8			3			4		
			5	4	9	7		
		3	7				1	
2								

Fiendish

								2
		2	4	9				6
8			1	3				
	8						4	7
			6		1			
4	5						2	
				2	3			5
1				6	5	9		
5								

Fiendish

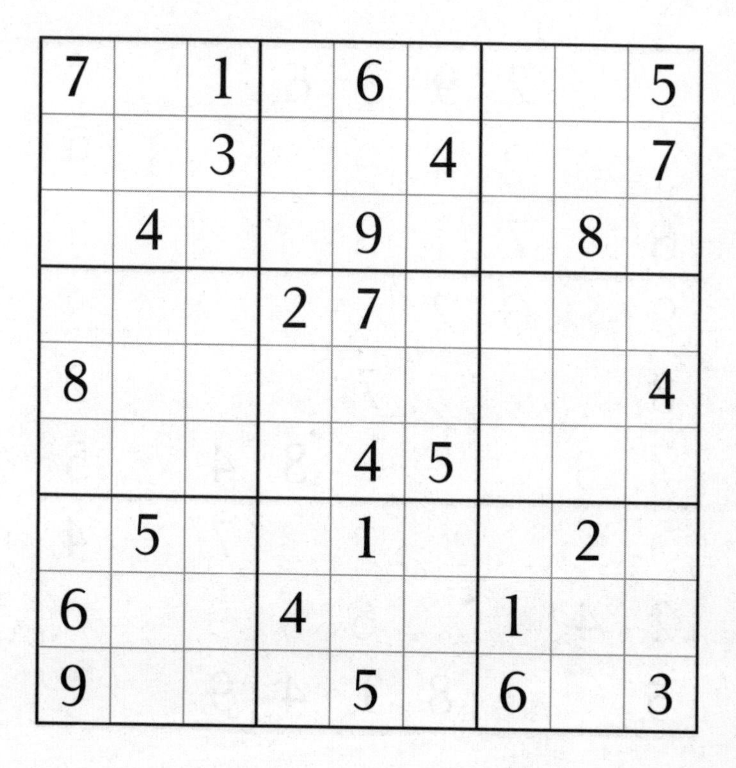

7		1		6				5
		3			4			7
	4			9			8	
			2	7				
8								4
				4	5			
	5			1			2	
6			4			1		
9				5		6		3

Fiendish

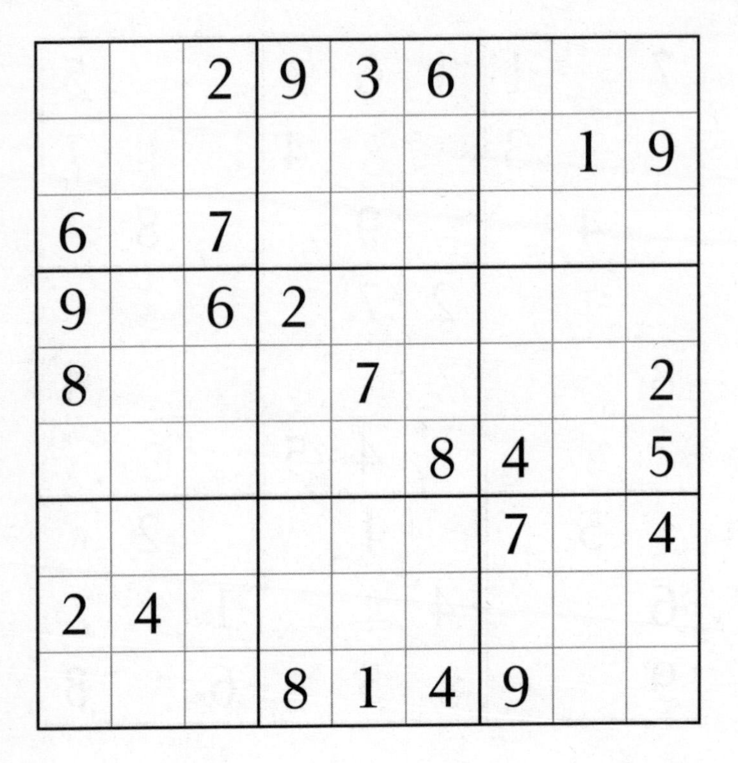

		2	9	3	6			
							1	9
6		7						
9		6	2					
8				7				2
					8	4		5
						7		4
2	4							
			8	1	4	9		

Fiendish

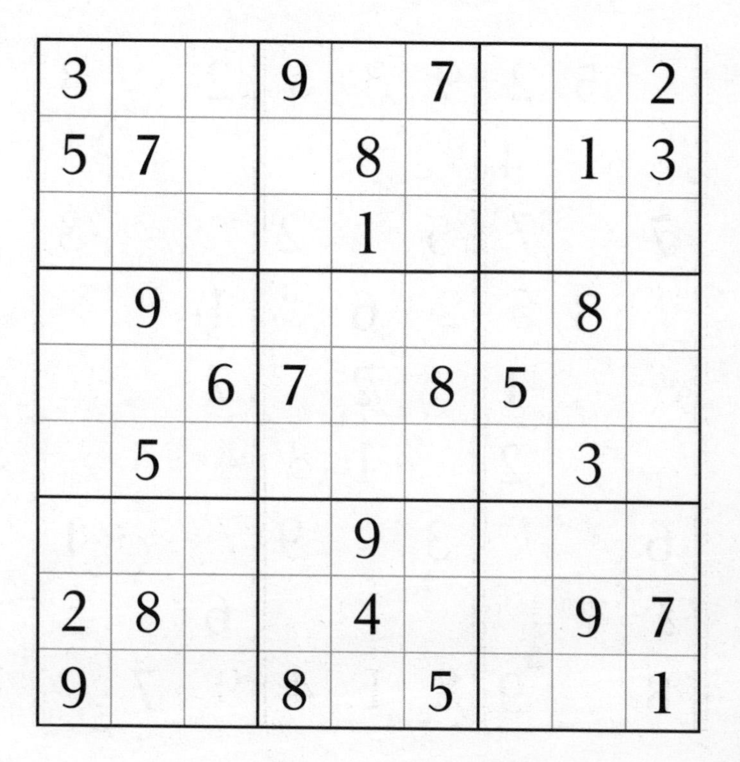

3			9		7			2
5	7			8			1	3
				1				
	9						8	
		6	7		8	5		
	5						3	
				9				
2	8			4			9	7
9			8		5			1

Fiendish

164

	5					2		3
		4						
7			5		2			8
	3			6		1		
		1		2		7		
		2		1			5	
6			3		9			1
						6		
8		9					7	

Fiendish

7						6		2
		2		3				5
4			8					1
					7		1	
		8	6		9	3		
	1		5					
8					4			9
5				6		7		
6		4						8

Fiendish

166

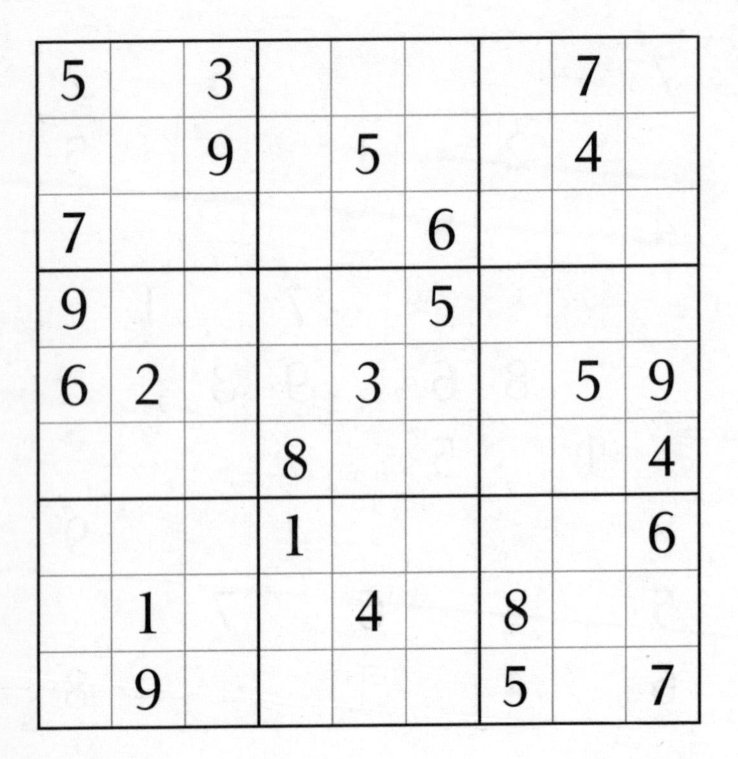

5		3					7	
		9		5			4	
7					6			
9					5			
6	2			3			5	9
			8					4
			1					6
	1			4		8		
	9					5		7

Fiendish

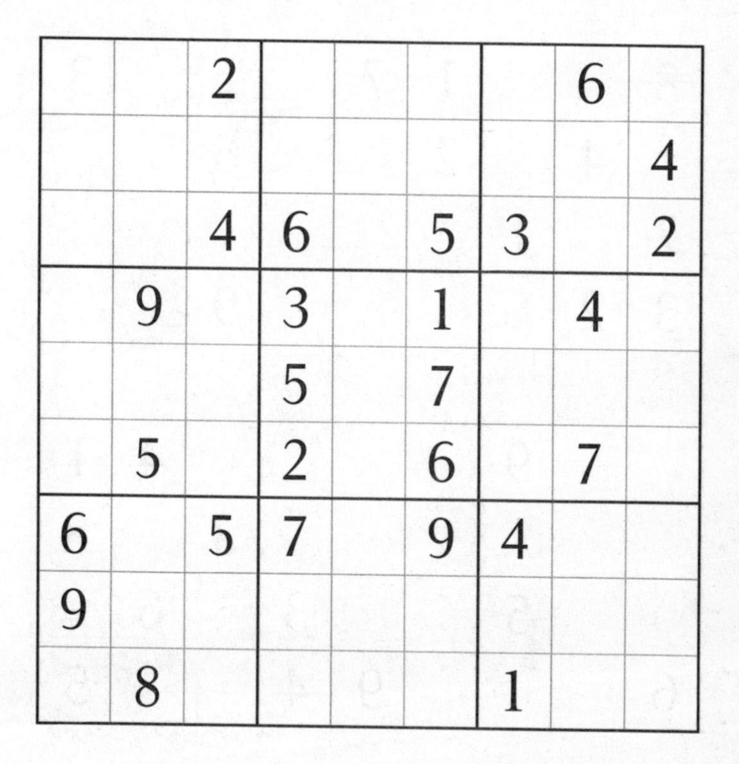

		2					6	
								4
		4	6		5	3		2
	9		3		1		4	
			5		7			
	5		2		6		7	
6		5	7		9	4		
9								
	8					1		

Fiendish

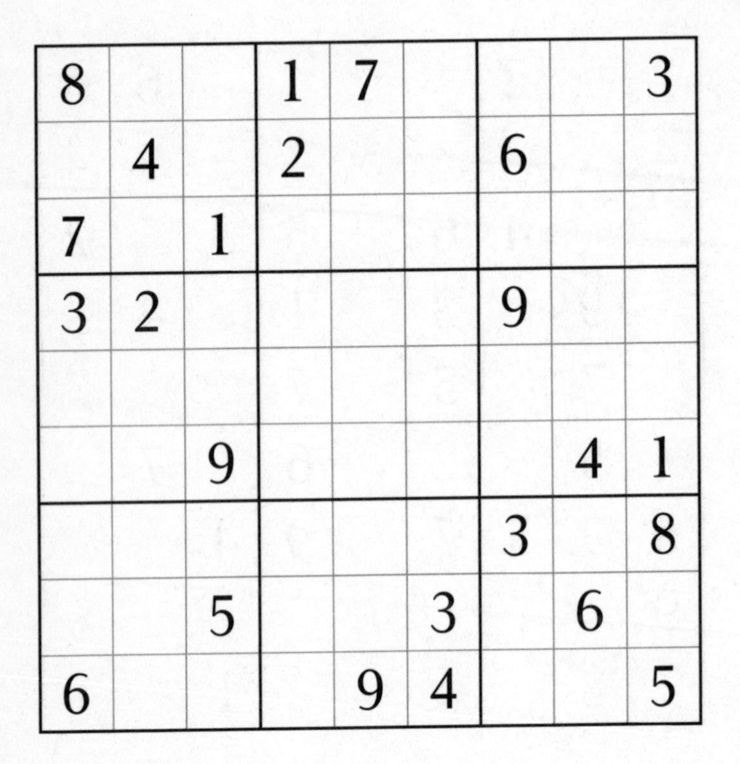

8			1	7				3
	4		2			6		
7		1						
3	2					9		
		9					4	1
						3		8
		5			3		6	
6				9	4			5

Fiendish

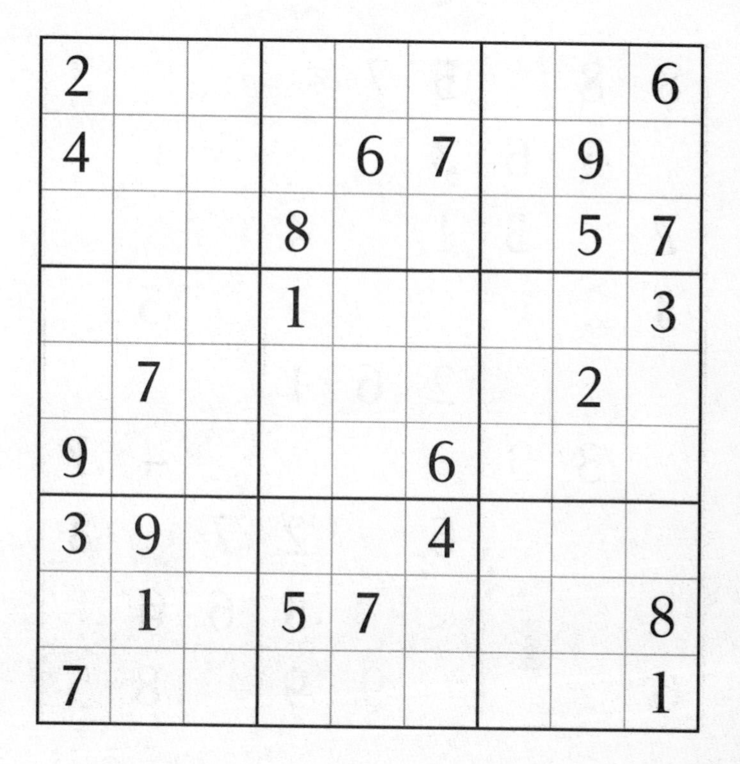

2								6
4				6	7		9	
			8				5	7
			1					3
	7						2	
9					6			
3	9				4			
	1		5	7				8
7								1

Fiendish

	8		5					
	5	6	3					
2		3	1		4			
4							5	
			2	6	1			
	3							7
			6		2	7		1
					5	6	4	
					9		8	

Fiendish

171

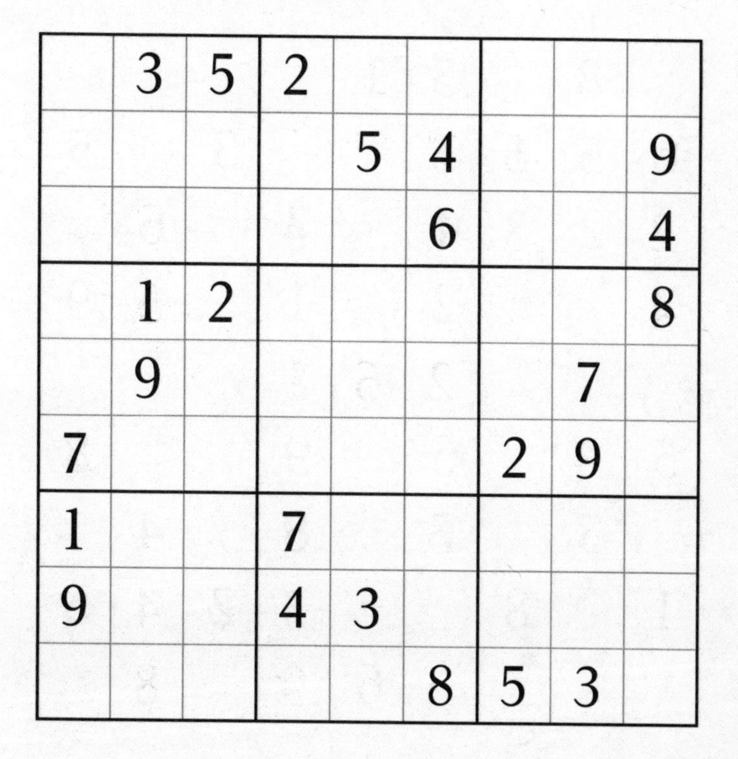

	3	5	2					
				5	4			9
					6			4
	1	2						8
	9						7	
7						2	9	
1			7					
9			4	3				
					8	5	3	

Fiendish

172

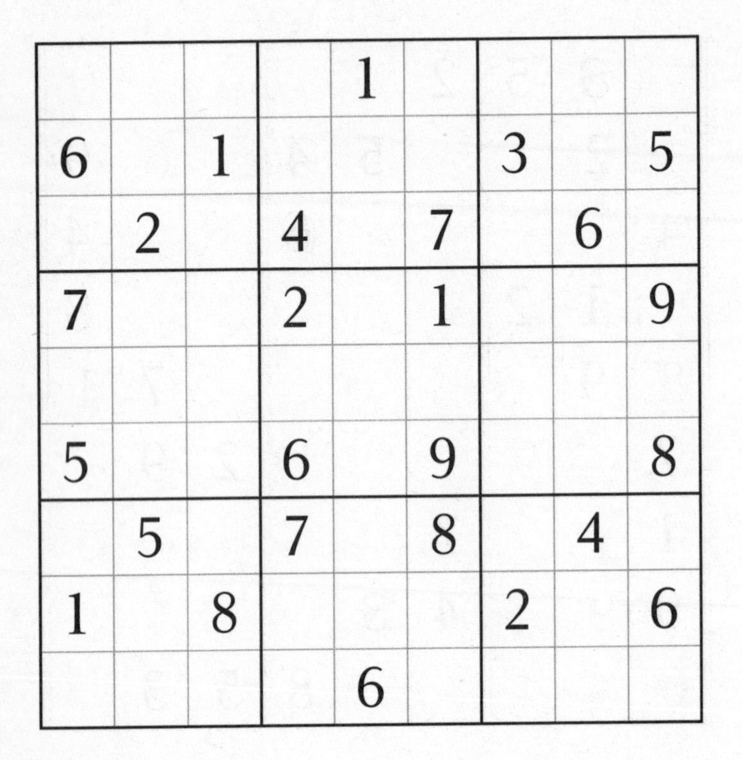

				1				
6		1				3		5
	2		4		7		6	
7			2		1			9
5			6		9			8
	5		7		8		4	
1		8				2		6
				6				

Fiendish

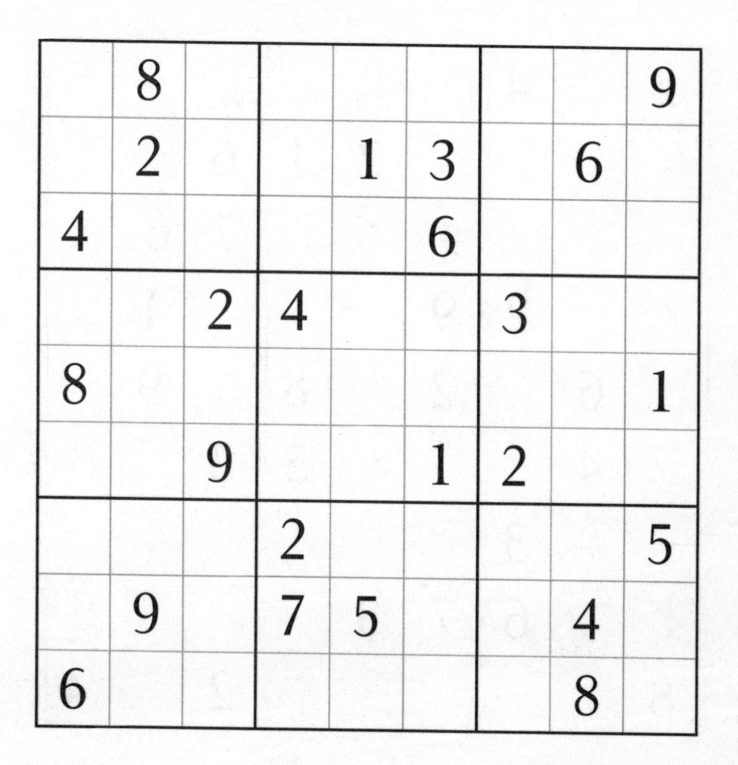

	8							9
	2			1	3		6	
4					6			
		2	4			3		
8								1
		9			1	2		
			2					5
	9		7	5			4	
6							8	

Fiendish

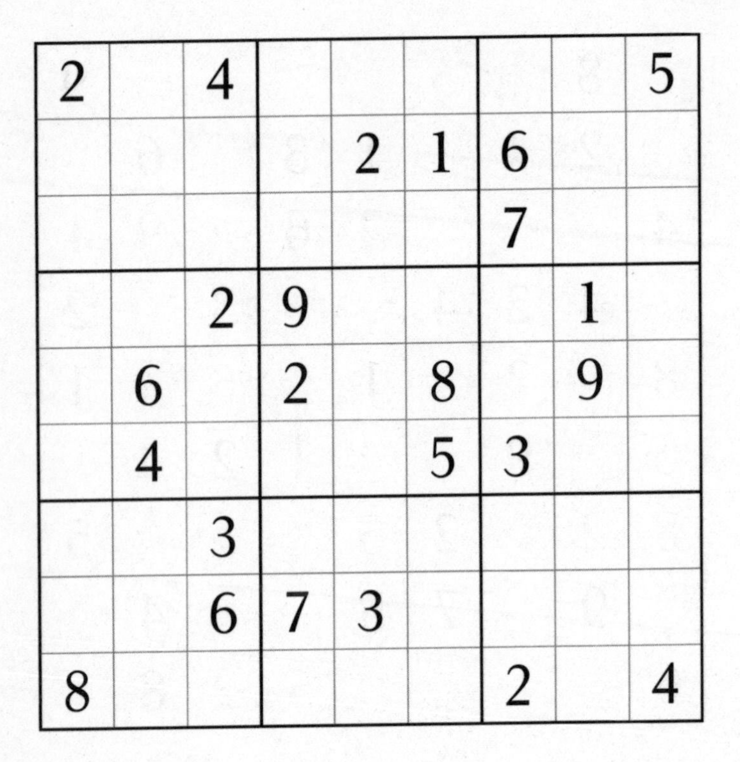

2		4						5
				2	1	6		
						7		
		2	9				1	
	6		2		8		9	
	4				5	3		
		3						
		6	7	3				
8						2		4

Fiendish

175

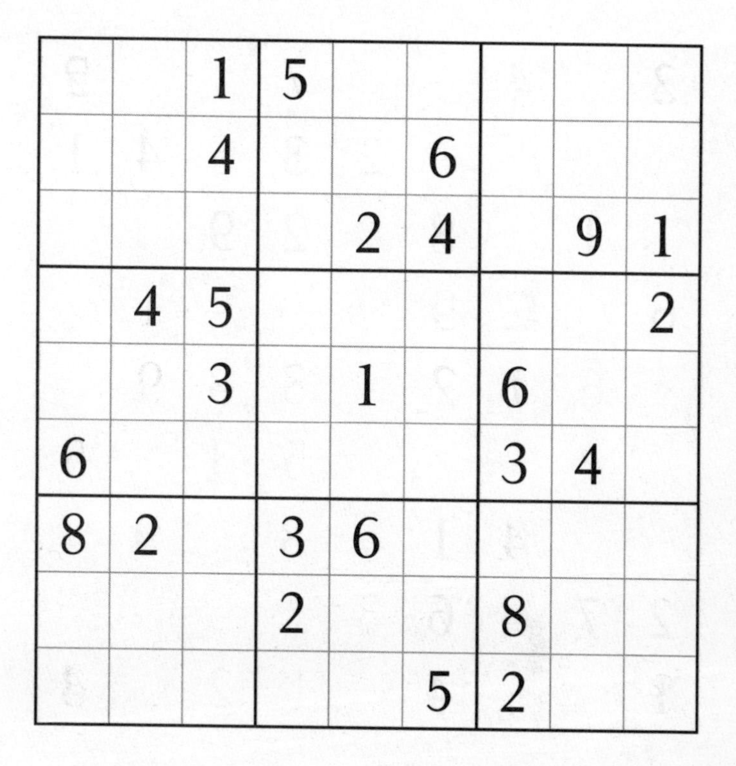

		1	5					
		4			6			
				2	4		9	1
	4	5						2
		3		1		6		
6						3	4	
8	2		3	6				
			2			8		
					5	2		

Fiendish

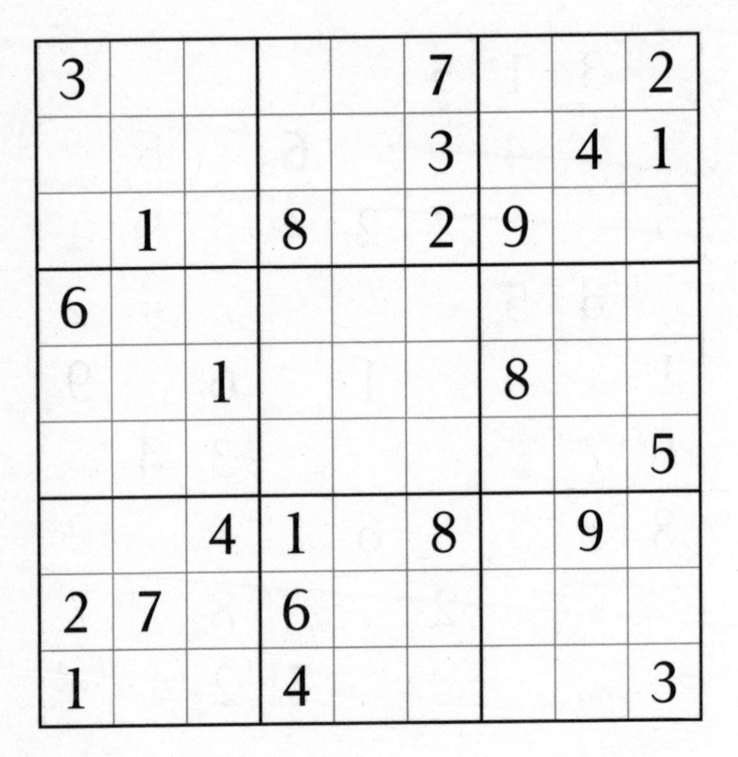

3					7			2
					3		4	1
	1		8		2	9		
6								
		1				8		
								5
		4	1		8		9	
2	7		6					
1			4					3

Fiendish

177

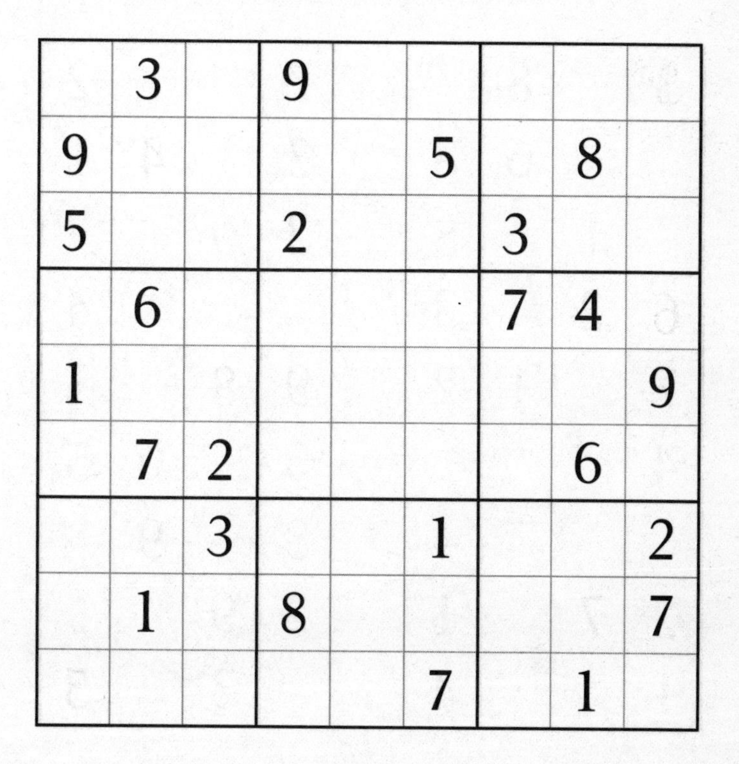

3		9					
9					5		8	
5			2			3		
	6					7	4	
1								9
	7	2					6	
		3			1			2
	1		8					7
					7		1	

Fiendish

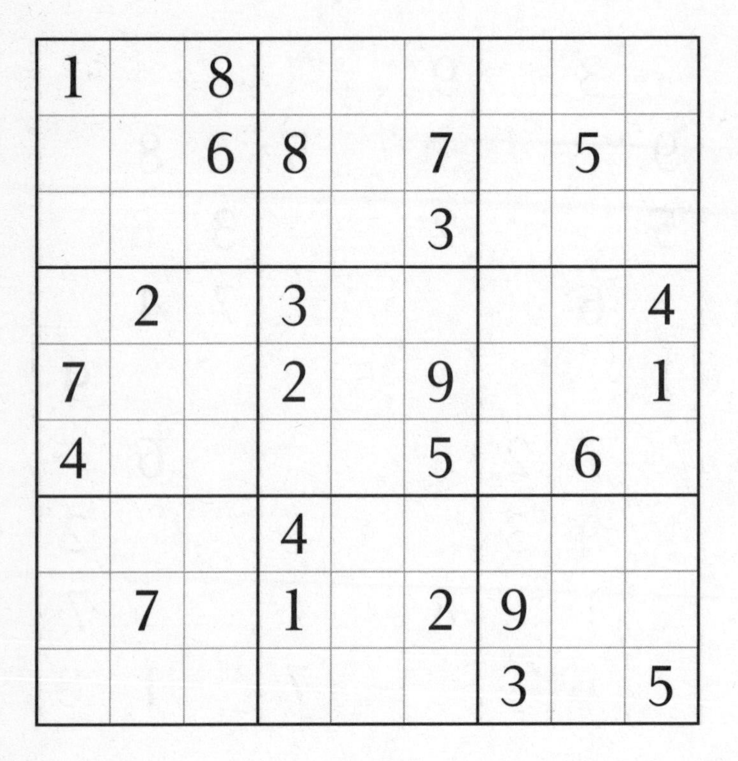

1		8						
		6	8		7		5	
					3			
	2		3					4
7			2		9			1
4					5		6	
			4					
	7		1		2	9		
						3		5

Fiendish

7					1		4	
	3		7			2		
9			4				8	
	4				8		1	
				5				
	5		1				2	
	1				5			3
		8			6		5	
	6		9					8

Fiendish

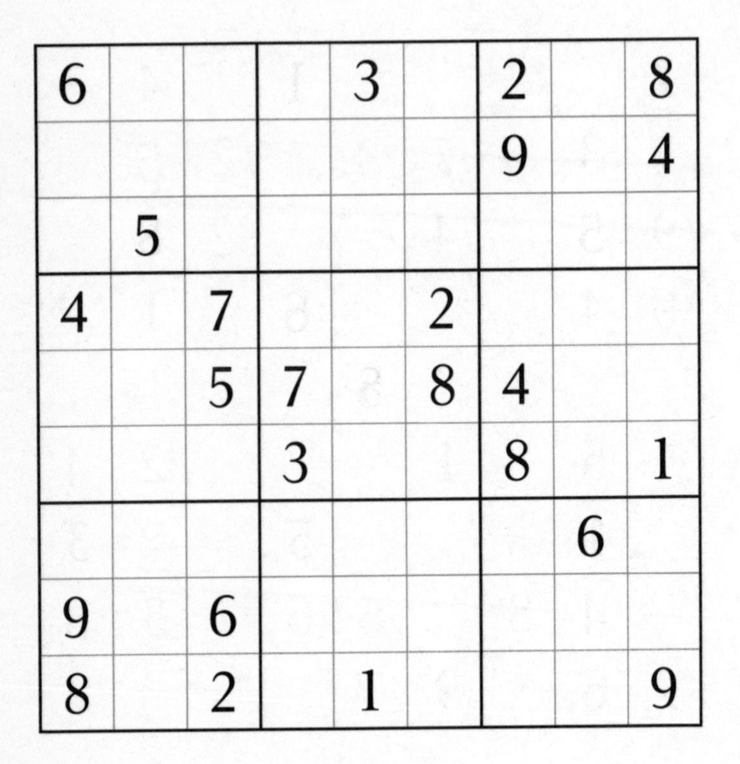

6				3		2		8
						9		4
	5							
4		7			2			
		5	7		8	4		
			3			8		1
							6	
9		6						
8		2		1				9

Fiendish

181

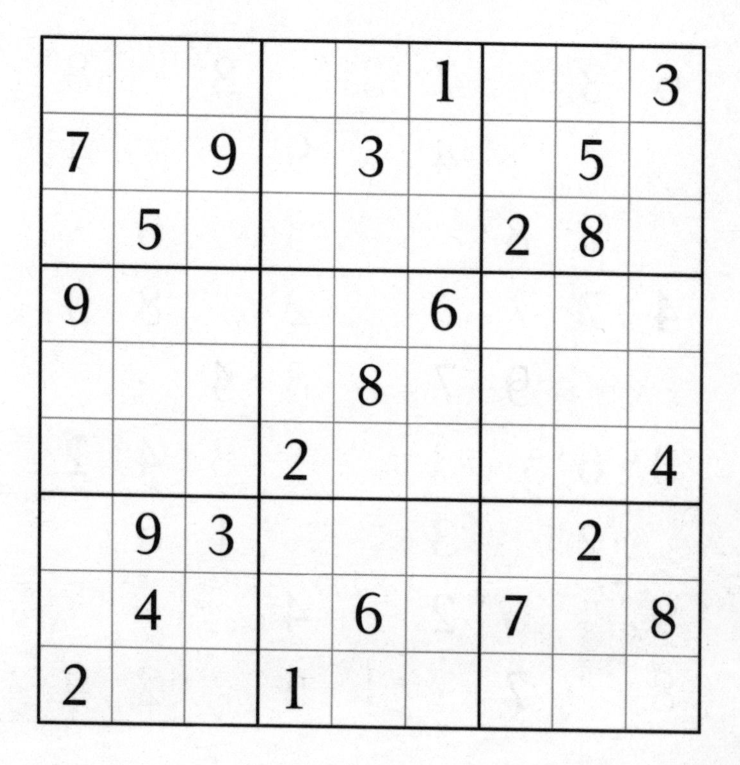

					1			3
7		9		3			5	
	5					2	8	
9					6			
				8				
			2					4
	9	3					2	
	4			6		7		8
2			1					

Fiendish

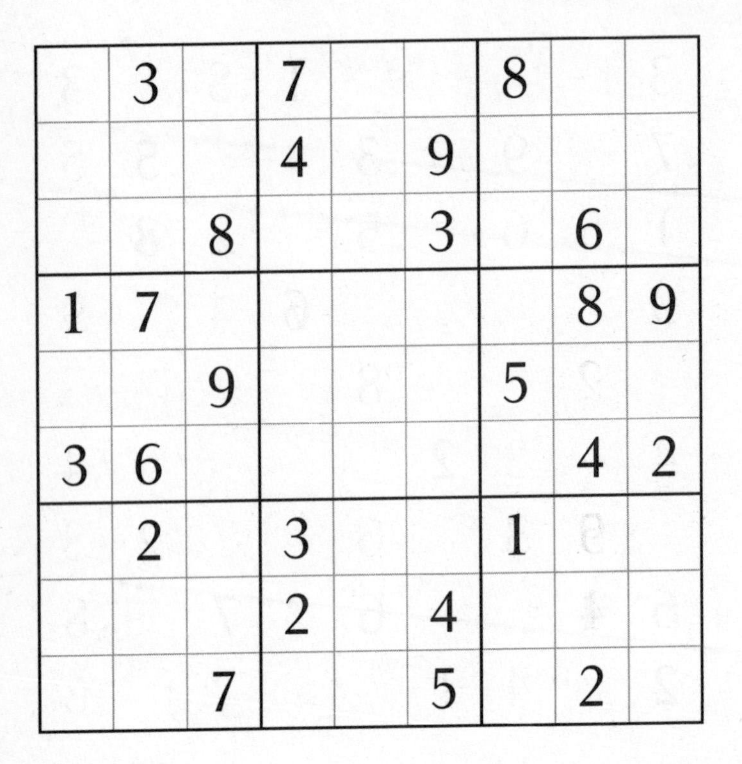

	3		7			8		
			4		9			
		8			3		6	
1	7						8	9
		9				5		
3	6						4	2
	2		3			1		
			2		4			
		7			5		2	

Fiendish

183

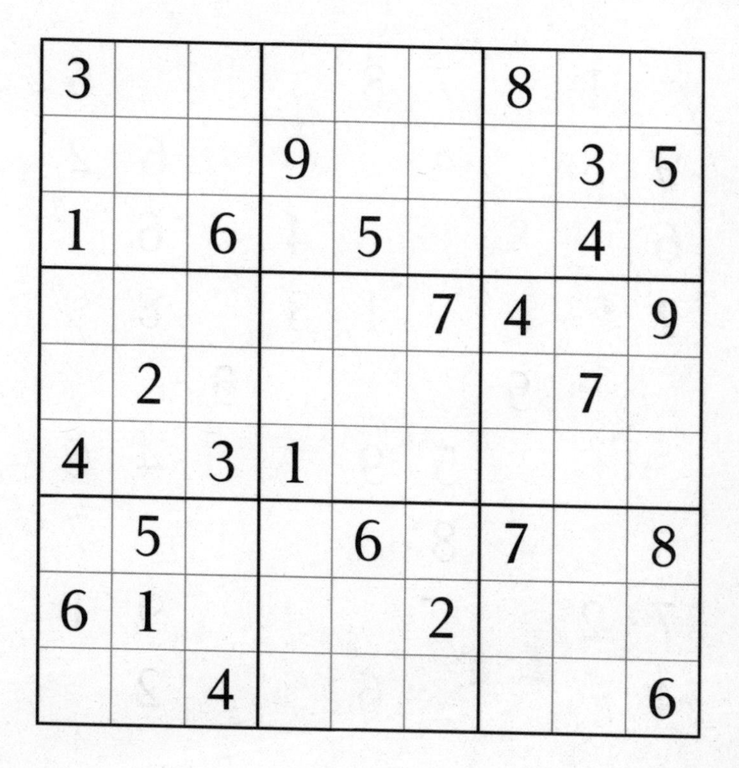

3						8		
			9				3	5
1		6		5			4	
					7	4		9
	2						7	
4		3	1					
	5			6		7		8
6	1				2			
		4						6

Fiendish

184

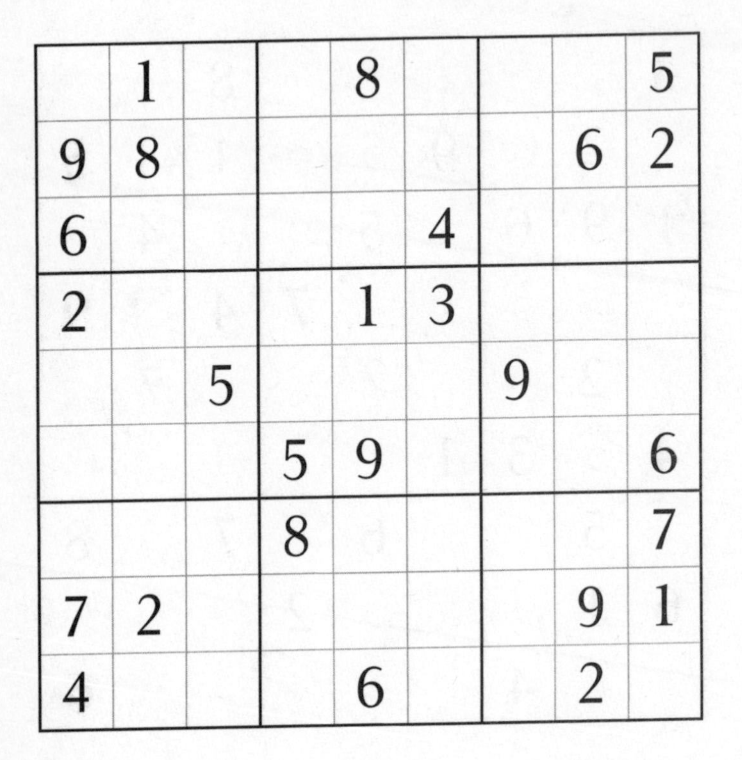

	1			8				5
9	8						6	2
6					4			
2				1	3			
		5				9		
			5	9				6
			8					7
7	2						9	1
4				6			2	

Fiendish

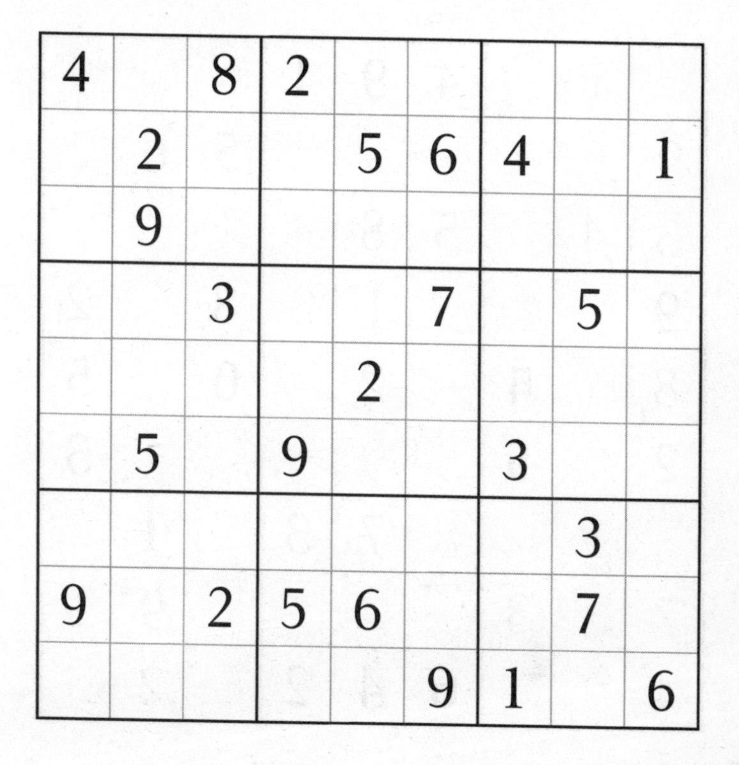

4		8	2					
	2			5	6	4		1
	9							
		3			7		5	
				2				
	5		9			3		
							3	
9		2	5	6			7	
					9	1		6

Fiendish

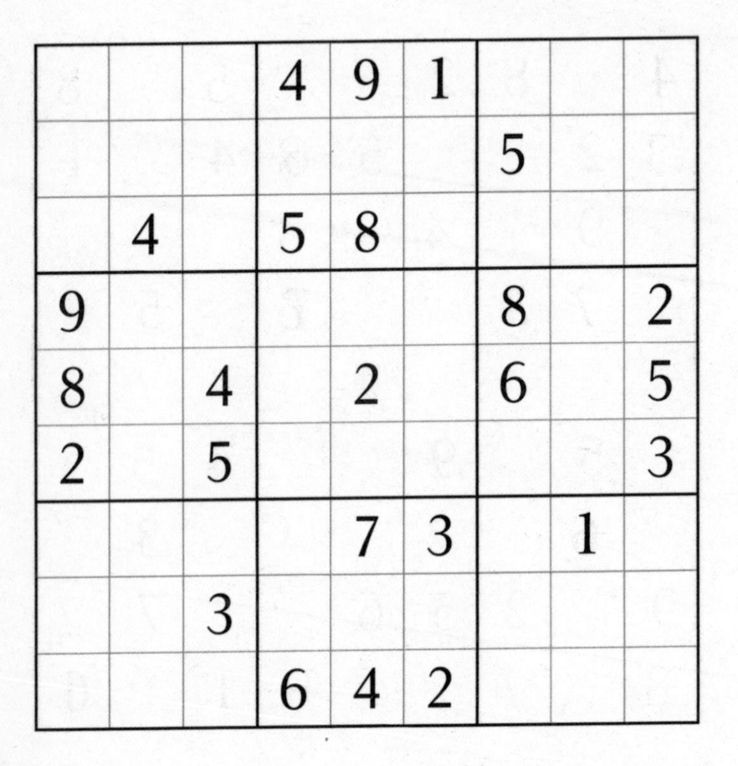

			4	9	1			
						5		
	4		5	8				
9						8		2
8		4		2		6		5
2		5						3
				7	3		1	
		3						
			6	4	2			

Fiendish

187

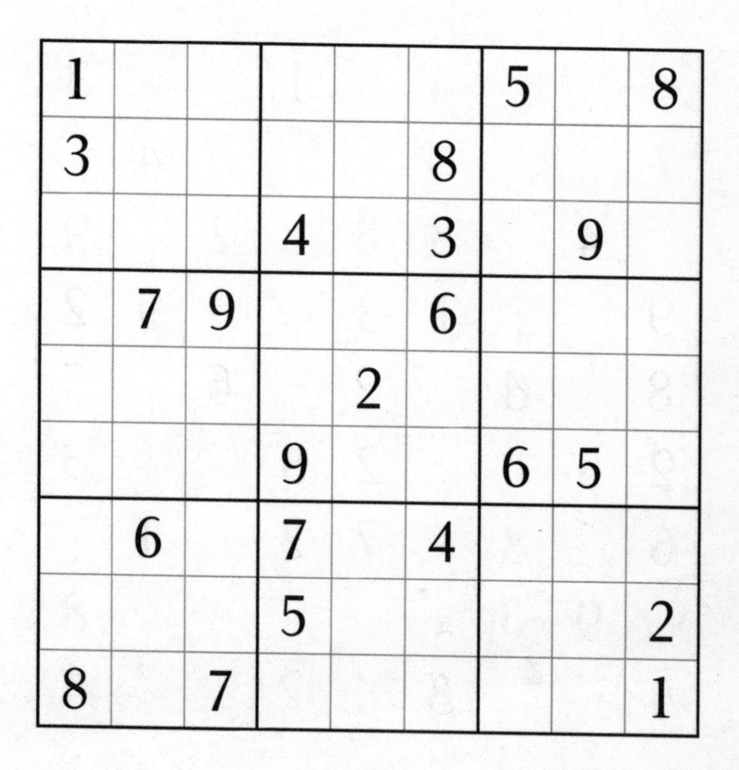

1						5		8
3					8			
			4		3		9	
	7	9			6			
				2				
			9			6	5	
	6		7		4			
			5					2
8		7						1

Fiendish

188

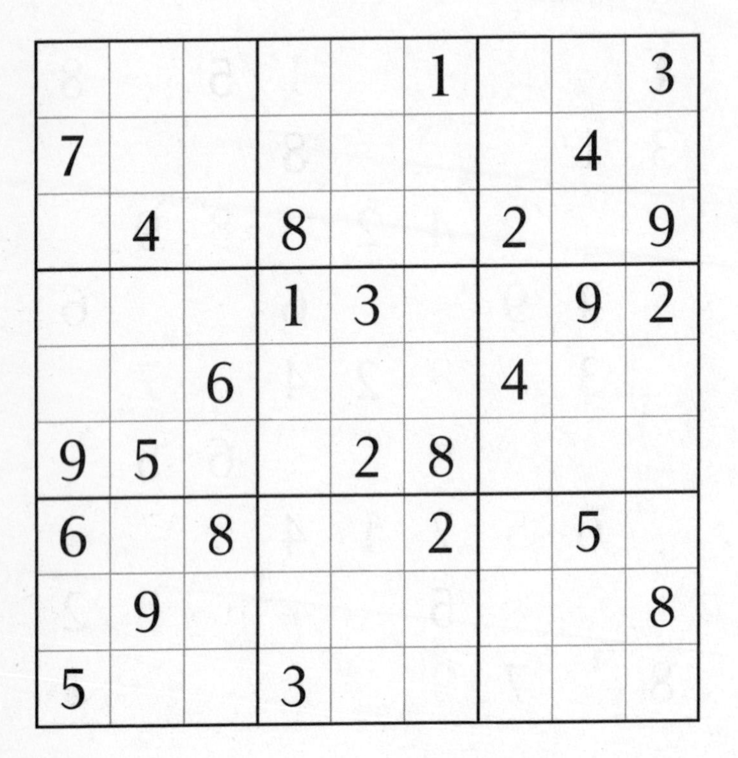

					1			3
7							4	
	4		8			2		9
			1	3			9	2
		6				4		
9	5			2	8			
6		8			2		5	
	9							8
5			3					

Fiendish

189

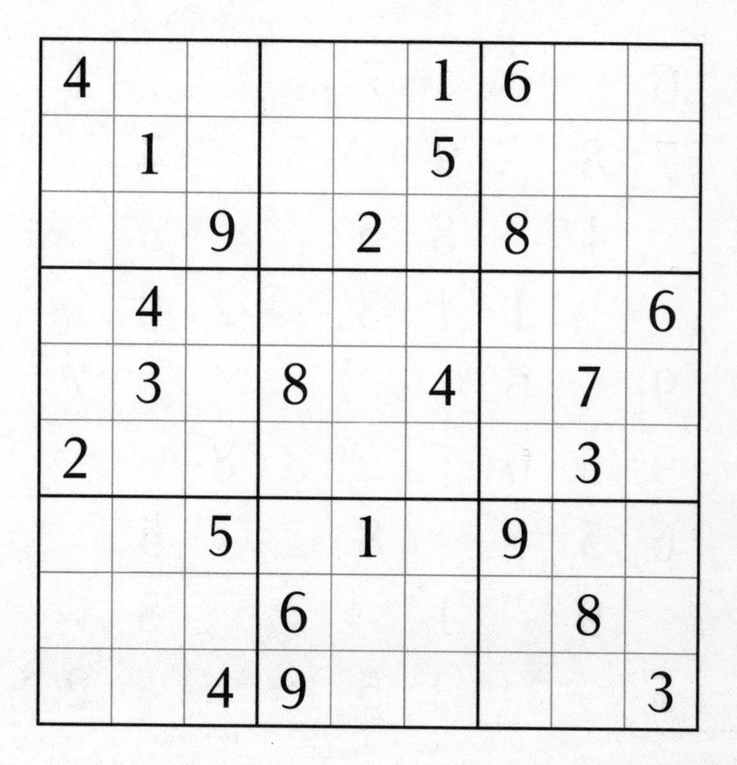

4					1	6		
	1				5			
		9		2		8		
	4							6
	3		8		4		7	
2							3	
		5		1		9		
			6				8	
		4	9					3

Fiendish

190

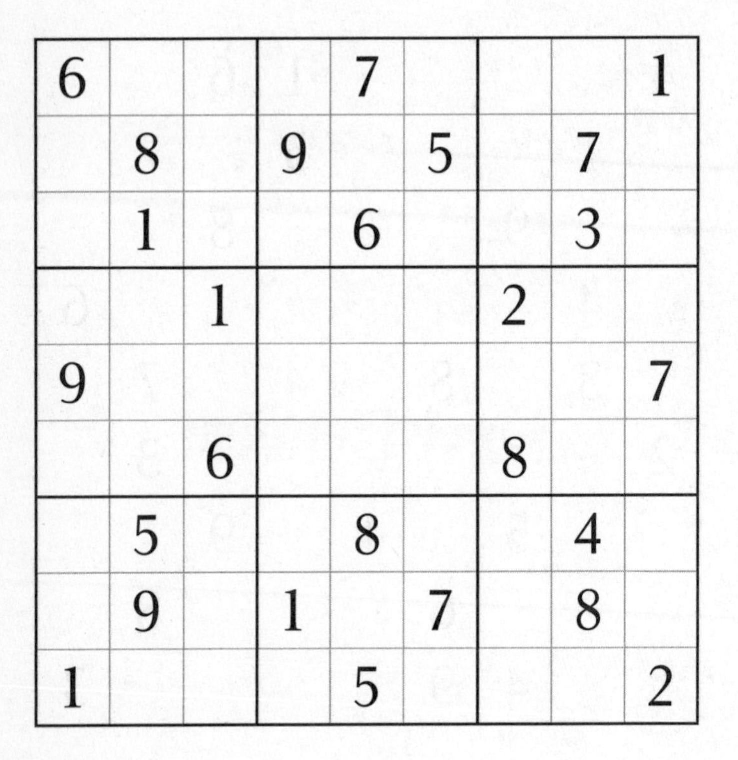

6				7				1
	8		9		5		7	
	1			6			3	
		1				2		
9								7
		6				8		
	5			8			4	
	9		1		7		8	
1				5				2

Fiendish

Super Fiendish

7					6			5
			3			9		
9		3				1	2	
					9	5		
		2				8		
		5	8					
	6	9				2		4
		8			1			
1			4					3

Super Fiendish

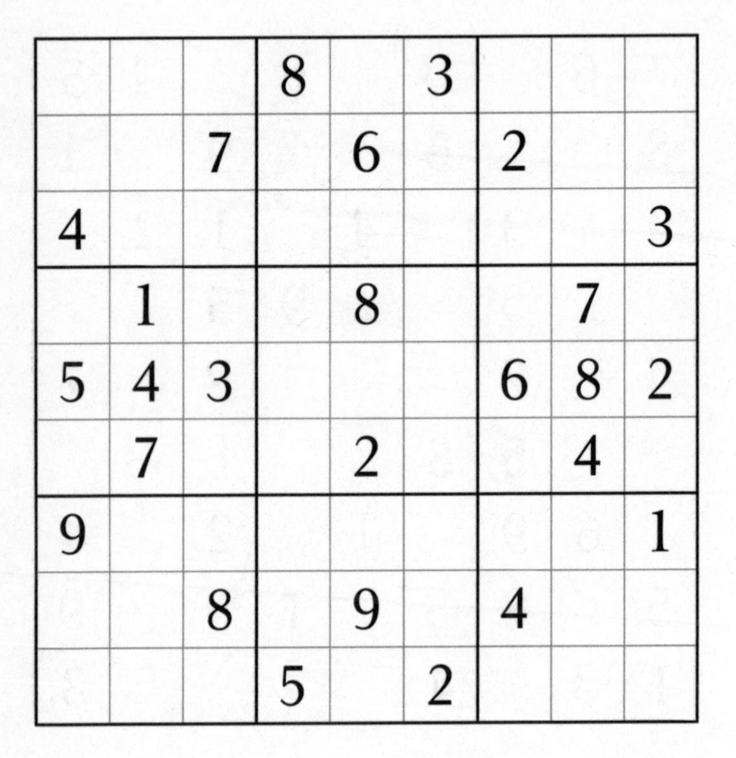

			8		3			
		7		6		2		
4								3
	1			8			7	
5	4	3				6	8	2
	7			2			4	
9								1
		8		9		4		
			5		2			

Super Fiendish

	6		9		8		4	
8			6		7			1
				4				
		5		2		4		
4		9				2		7
		6		7		8		
				1				
5			8		3			9
	3		7		2		5	

Super Fiendish

		9		6		3		
		6	8	4		7		
8			9					
				7			4	
3		4				2		1
	9			5				
					6			5
		2		3	5	1		
		7		2		8		

Super Fiendish

195

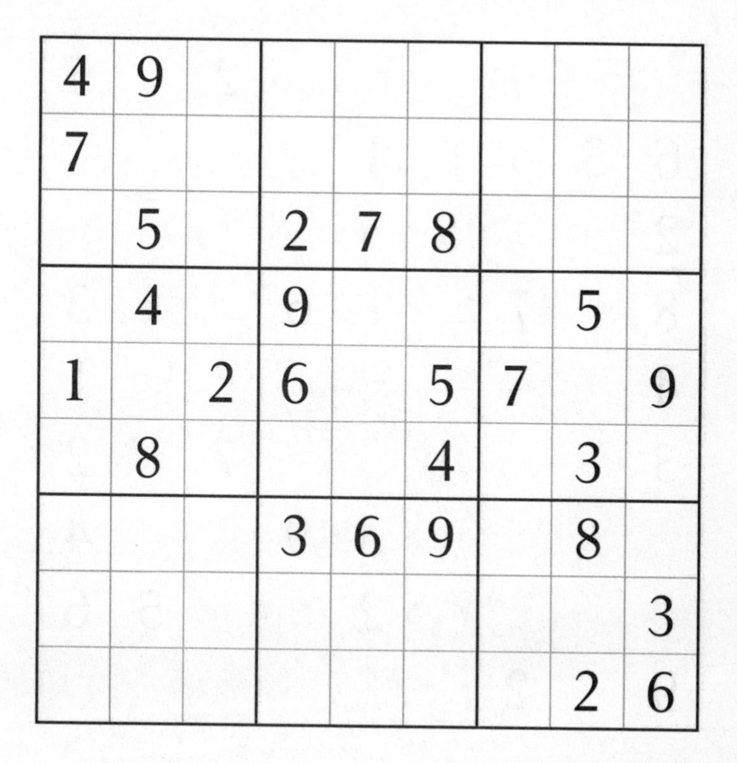

4	9							
7								
	5		2	7	8			
	4		9				5	
1		2	6		5	7		9
	8				4		3	
			3	6	9		8	
								3
							2	6

Super Fiendish

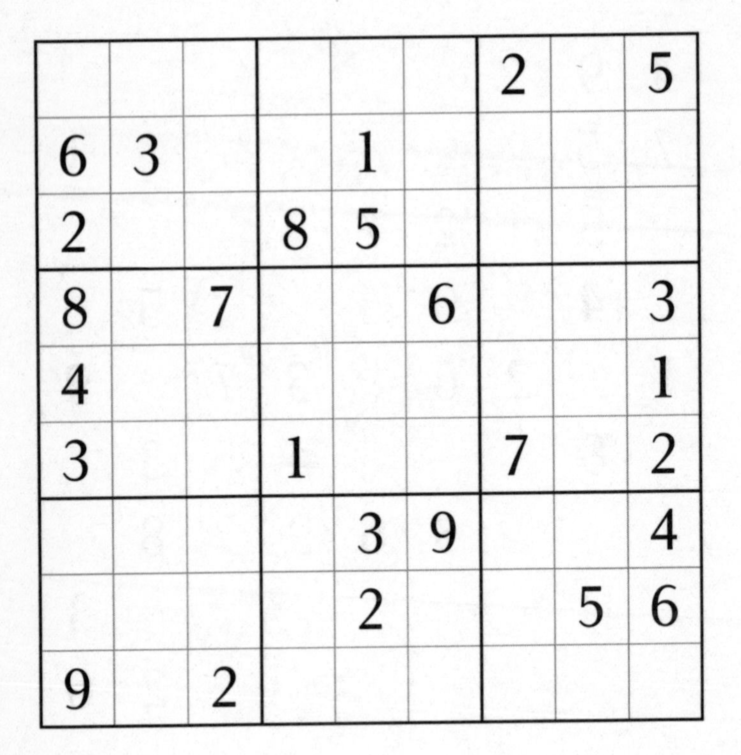

						2		5
6	3			1				
2			8	5				
8		7			6			3
4								1
3			1			7		2
				3	9			4
				2			5	6
9		2						

Super Fiendish

2								
	5	7					4	1
		1		8				
	2	4	3	5				
8			7		2			3
				4	6	5	2	
				3		7		
9	4					2	6	
								5

Super Fiendish

198

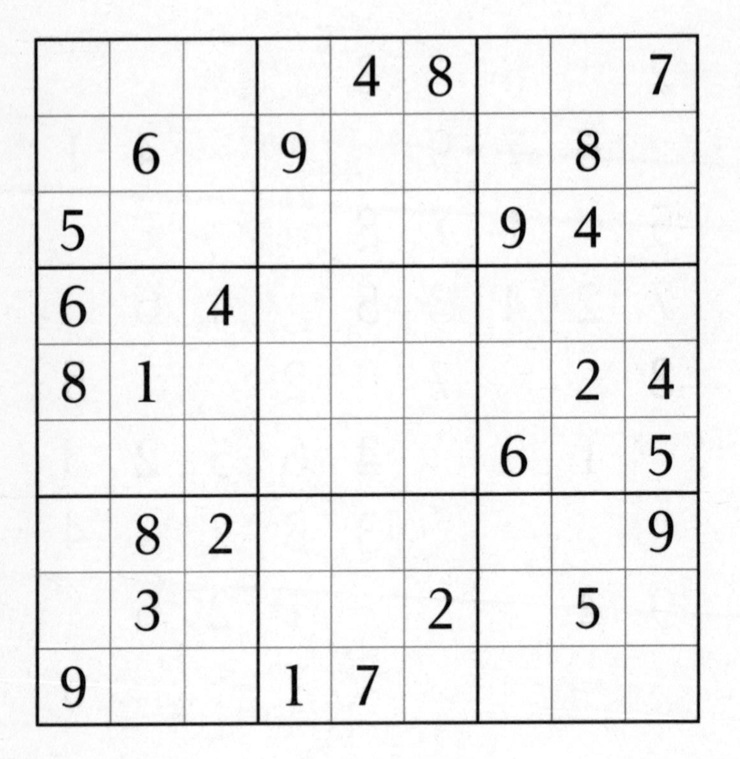

				4	8			7
	6		9				8	
5						9	4	
6		4						
8	1						2	4
						6		5
	8	2						9
	3				2		5	
9			1	7				

Super Fiendish

		8		3				
	6		9				4	
5			7					
7				8		4	6	
3								5
	1	9		2				3
					3			4
	2				1		8	
				4		1		

Super Fiendish

200

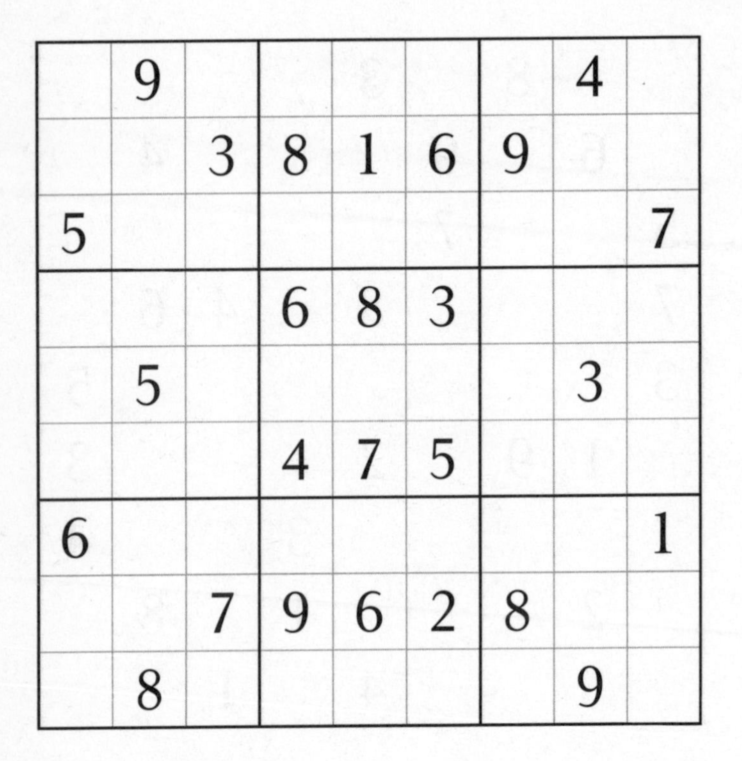

	9						4	
		3	8	1	6	9		
5								7
			6	8	3			
	5						3	
			4	7	5			
6								1
		7	9	6	2	8		
	8						9	

Super Fiendish

Solutions

1

3	4	9	1	6	7	5	2	8
5	7	8	2	3	4	9	6	1
1	6	2	9	5	8	7	3	4
6	2	5	4	7	3	1	8	9
8	3	7	6	9	1	2	4	5
4	9	1	8	2	5	3	7	6
2	5	6	3	4	9	8	1	7
9	1	3	7	8	6	4	5	2
7	8	4	5	1	2	6	9	3

2

7	4	8	2	3	5	9	1	6
6	1	2	7	4	9	5	8	3
9	5	3	6	1	8	7	2	4
3	6	4	5	2	1	8	9	7
8	7	1	4	9	3	2	6	5
2	9	5	8	6	7	3	4	1
1	8	9	3	5	6	4	7	2
4	3	7	1	8	2	6	5	9
5	2	6	9	7	4	1	3	8

3

3	4	2	6	1	7	9	8	5
7	1	8	2	5	9	3	4	6
9	6	5	4	8	3	7	2	1
8	7	4	9	2	6	5	1	3
2	3	9	5	4	1	6	7	8
6	5	1	3	7	8	4	9	2
5	2	6	1	9	4	8	3	7
1	9	7	8	3	5	2	6	4
4	8	3	7	6	2	1	5	9

4

4	9	8	3	6	1	5	2	7
6	1	5	7	8	2	3	9	4
2	7	3	5	9	4	1	8	6
8	3	4	6	7	9	2	5	1
1	2	9	4	3	5	7	6	8
7	5	6	2	1	8	9	4	3
3	8	7	9	5	6	4	1	2
9	6	2	1	4	3	8	7	5
5	4	1	8	2	7	6	3	9

5

7	5	2	8	6	9	3	4	1
9	1	6	5	4	3	2	7	8
3	4	8	2	7	1	6	5	9
4	9	1	6	8	7	5	2	3
6	2	5	9	3	4	1	8	7
8	3	7	1	5	2	9	6	4
5	6	9	7	1	8	4	3	2
2	8	3	4	9	6	7	1	5
1	7	4	3	2	5	8	9	6

6

6	4	3	8	5	2	7	1	9
8	9	1	6	3	7	5	2	4
7	5	2	4	1	9	6	3	8
2	1	4	7	6	8	9	5	3
9	7	8	3	2	5	4	6	1
3	6	5	1	9	4	2	8	7
1	3	9	5	7	6	8	4	2
5	8	7	2	4	3	1	9	6
4	2	6	9	8	1	3	7	5

7

3	1	8	2	5	4	9	7	6
6	2	7	9	8	3	4	5	1
5	4	9	6	7	1	3	2	8
4	9	5	1	6	7	2	8	3
1	7	2	3	4	8	6	9	5
8	6	3	5	2	9	1	4	7
2	5	4	8	3	6	7	1	9
9	8	6	7	1	2	5	3	4
7	3	1	4	9	5	8	6	2

8

9	1	4	5	8	7	3	6	2
3	7	8	1	6	2	9	4	5
2	5	6	4	9	3	7	8	1
7	8	9	3	2	1	6	5	4
1	6	3	9	4	5	8	2	7
5	4	2	8	7	6	1	3	9
6	9	5	2	1	8	4	7	3
4	3	7	6	5	9	2	1	8
8	2	1	7	3	4	5	9	6

9

7	8	2	3	1	4	5	9	6
1	5	9	6	7	8	3	4	2
6	4	3	5	2	9	8	7	1
4	1	6	8	5	7	9	2	3
9	3	5	4	6	2	1	8	7
8	2	7	9	3	1	6	5	4
3	9	1	2	4	5	7	6	8
2	6	8	7	9	3	4	1	5
5	7	4	1	8	6	2	3	9

10

6	7	5	1	4	8	9	2	3
2	4	8	5	9	3	1	6	7
1	9	3	2	7	6	5	8	4
8	5	2	3	6	1	4	7	9
4	1	9	7	8	2	6	3	5
3	6	7	4	5	9	2	1	8
7	2	1	9	3	4	8	5	6
9	3	6	8	1	5	7	4	2
5	8	4	6	2	7	3	9	1

11

6	3	8	4	9	5	2	7	1
4	7	1	8	2	6	5	9	3
2	5	9	7	1	3	8	4	6
3	8	7	5	4	2	1	6	9
9	4	2	3	6	1	7	8	5
1	6	5	9	7	8	4	3	2
8	1	6	2	3	4	9	5	7
7	2	4	6	5	9	3	1	8
5	9	3	1	8	7	6	2	4

12

8	9	3	4	2	7	5	6	1
5	6	1	8	9	3	7	2	4
7	2	4	5	6	1	3	9	8
3	1	7	2	8	4	6	5	9
9	8	2	3	5	6	1	4	7
6	4	5	1	7	9	8	3	2
4	7	6	9	1	5	2	8	3
2	5	9	7	3	8	4	1	6
1	3	8	6	4	2	9	7	5

13

8	4	6	9	7	3	5	2	1
7	9	3	2	5	1	4	8	6
2	5	1	6	8	4	7	3	9
3	7	9	4	2	5	1	6	8
6	2	5	7	1	8	3	9	4
4	1	8	3	6	9	2	7	5
1	3	2	5	9	6	8	4	7
5	6	7	8	4	2	9	1	3
9	8	4	1	3	7	6	5	2

14

1	6	5	2	8	9	7	3	4
2	9	4	1	3	7	5	6	8
8	7	3	5	4	6	1	9	2
3	1	9	8	5	4	6	2	7
6	5	7	9	1	2	8	4	3
4	8	2	7	6	3	9	5	1
5	4	1	6	2	8	3	7	9
9	3	6	4	7	1	2	8	5
7	2	8	3	9	5	4	1	6

15

3	7	5	4	6	9	8	2	1
2	1	9	7	5	8	3	6	4
8	6	4	2	3	1	9	5	7
1	8	2	6	4	3	5	7	9
4	5	6	8	9	7	2	1	3
7	9	3	1	2	5	4	8	6
5	4	1	9	8	6	7	3	2
9	3	7	5	1	2	6	4	8
6	2	8	3	7	4	1	9	5

16

7	8	5	2	4	1	3	6	9
6	4	2	8	3	9	7	1	5
9	1	3	6	7	5	4	2	8
3	7	1	9	8	6	2	5	4
5	9	8	4	1	2	6	7	3
2	6	4	3	5	7	9	8	1
4	5	6	7	9	8	1	3	2
1	3	7	5	2	4	8	9	6
8	2	9	1	6	3	5	4	7

17

1	7	6	4	3	5	9	8	2
4	2	5	8	7	9	6	3	1
9	8	3	6	2	1	5	7	4
6	9	7	2	4	8	3	1	5
8	1	4	7	5	3	2	9	6
3	5	2	1	9	6	7	4	8
7	6	9	5	8	4	1	2	3
5	3	8	9	1	2	4	6	7
2	4	1	3	6	7	8	5	9

18

3	1	2	7	6	8	4	5	9
5	6	9	4	3	1	7	2	8
7	8	4	9	5	2	1	3	6
6	3	8	5	9	7	2	4	1
1	9	5	8	2	4	3	6	7
4	2	7	3	1	6	9	8	5
9	4	6	1	8	3	5	7	2
8	7	1	2	4	5	6	9	3
2	5	3	6	7	9	8	1	4

19

7	2	3	9	5	4	1	8	6
5	8	4	6	2	1	9	3	7
9	1	6	8	3	7	5	2	4
2	7	8	3	4	9	6	1	5
3	6	9	2	1	5	7	4	8
1	4	5	7	6	8	2	9	3
6	9	7	4	8	2	3	5	1
4	5	2	1	7	3	8	6	9
8	3	1	5	9	6	4	7	2

20

1	8	7	6	5	9	3	4	2
5	2	6	1	3	4	9	8	7
3	4	9	2	8	7	1	6	5
7	9	3	4	6	1	2	5	8
4	1	2	5	7	8	6	9	3
8	6	5	9	2	3	7	1	4
9	7	4	8	1	2	5	3	6
6	3	8	7	9	5	4	2	1
2	5	1	3	4	6	8	7	9

21

5	8	1	6	3	7	4	9	2
9	6	4	5	2	8	1	7	3
2	7	3	1	9	4	5	6	8
3	9	7	2	6	5	8	1	4
4	5	2	7	8	1	6	3	9
8	1	6	9	4	3	7	2	5
7	3	5	4	1	9	2	8	6
6	4	9	8	7	2	3	5	1
1	2	8	3	5	6	9	4	7

22

1	6	8	5	3	9	7	2	4
9	5	4	6	7	2	8	3	1
2	3	7	1	8	4	5	6	9
7	8	3	9	5	1	6	4	2
6	1	5	2	4	8	3	9	7
4	9	2	3	6	7	1	5	8
8	2	6	4	1	5	9	7	3
5	4	1	7	9	3	2	8	6
3	7	9	8	2	6	4	1	5

23

7	8	5	4	3	1	6	2	9
9	1	4	2	8	6	5	3	7
6	3	2	5	9	7	8	1	4
5	6	1	8	4	2	7	9	3
4	7	9	6	1	3	2	8	5
3	2	8	7	5	9	4	6	1
2	4	3	9	7	8	1	5	6
1	5	6	3	2	4	9	7	8
8	9	7	1	6	5	3	4	2

24

9	2	6	5	4	3	8	1	7
8	1	7	2	9	6	3	5	4
4	3	5	7	8	1	9	2	6
5	4	9	8	2	7	6	3	1
2	8	1	3	6	4	7	9	5
7	6	3	1	5	9	4	8	2
6	5	8	9	7	2	1	4	3
1	7	2	4	3	8	5	6	9
3	9	4	6	1	5	2	7	8

25

6	4	1	5	8	3	7	2	9
9	2	5	7	4	6	1	8	3
8	7	3	9	1	2	6	5	4
4	1	6	3	5	7	2	9	8
7	3	9	2	6	8	4	1	5
2	5	8	1	9	4	3	6	7
3	6	2	8	7	5	9	4	1
5	9	7	4	2	1	8	3	6
1	8	4	6	3	9	5	7	2

26

6	9	1	4	5	7	2	8	3
3	7	4	8	9	2	6	5	1
2	5	8	6	1	3	7	4	9
4	8	3	1	2	9	5	7	6
5	6	2	7	3	4	9	1	8
7	1	9	5	6	8	4	3	2
1	4	6	9	8	5	3	2	7
9	3	7	2	4	1	8	6	5
8	2	5	3	7	6	1	9	4

27

7	4	6	1	3	2	5	8	9
3	8	9	4	6	5	2	7	1
2	1	5	7	9	8	4	3	6
6	5	2	3	8	7	1	9	4
8	9	3	2	4	1	6	5	7
1	7	4	9	5	6	8	2	3
4	2	7	5	1	3	9	6	8
5	6	1	8	7	9	3	4	2
9	3	8	6	2	4	7	1	5

28

2	9	3	1	5	4	8	7	6
8	1	4	7	9	6	3	2	5
6	5	7	8	2	3	9	1	4
7	4	6	3	1	2	5	9	8
9	8	2	6	7	5	4	3	1
5	3	1	4	8	9	2	6	7
3	7	8	9	4	1	6	5	2
4	2	9	5	6	7	1	8	3
1	6	5	2	3	8	7	4	9

29

9	8	4	3	2	5	6	1	7
5	6	7	9	8	1	3	4	2
2	1	3	7	4	6	8	5	9
3	2	5	8	1	7	9	6	4
6	9	1	4	3	2	7	8	5
7	4	8	5	6	9	2	3	1
1	7	6	2	5	3	4	9	8
8	3	9	1	7	4	5	2	6
4	5	2	6	9	8	1	7	3

30

5	4	3	6	8	7	1	2	9
9	7	8	1	5	2	3	4	6
2	6	1	9	3	4	5	8	7
4	8	5	7	2	6	9	3	1
3	9	6	5	1	8	4	7	2
7	1	2	3	4	9	8	6	5
6	5	7	8	9	3	2	1	4
8	2	9	4	6	1	7	5	3
1	3	4	2	7	5	6	9	8

31

6	7	5	3	4	2	9	8	1
8	4	3	6	1	9	2	5	7
1	2	9	7	5	8	4	6	3
5	9	6	4	3	1	8	7	2
3	8	7	2	9	5	1	4	6
2	1	4	8	7	6	5	3	9
9	6	8	5	2	7	3	1	4
7	3	1	9	8	4	6	2	5
4	5	2	1	6	3	7	9	8

32

9	5	8	2	6	4	1	7	3
4	1	7	5	8	3	6	9	2
3	2	6	7	1	9	4	5	8
6	3	1	8	5	2	9	4	7
7	4	9	1	3	6	8	2	5
5	8	2	4	9	7	3	1	6
8	6	5	9	7	1	2	3	4
2	9	3	6	4	5	7	8	1
1	7	4	3	2	8	5	6	9

33

2	4	3	8	5	1	7	9	6
7	6	9	4	3	2	1	8	5
8	1	5	6	9	7	4	2	3
6	7	4	9	1	5	8	3	2
9	5	1	2	8	3	6	4	7
3	2	8	7	6	4	5	1	9
4	3	6	5	2	8	9	7	1
5	8	2	1	7	9	3	6	4
1	9	7	3	4	6	2	5	8

34

4	1	6	2	7	8	5	3	9
7	5	2	1	9	3	4	6	8
3	9	8	5	6	4	2	7	1
8	4	5	9	3	6	7	1	2
2	6	9	4	1	7	3	8	5
1	3	7	8	2	5	6	9	4
5	8	3	7	4	1	9	2	6
9	7	4	6	8	2	1	5	3
6	2	1	3	5	9	8	4	7

35

7	3	4	1	5	6	8	9	2
8	6	9	4	7	2	1	3	5
2	1	5	3	9	8	6	7	4
1	7	8	9	4	5	2	6	3
9	5	3	2	6	1	7	4	8
6	4	2	8	3	7	9	5	1
5	9	1	7	8	3	4	2	6
3	2	7	6	1	4	5	8	9
4	8	6	5	2	9	3	1	7

36

8	2	4	1	9	3	7	6	5
5	7	9	8	4	6	2	3	1
6	3	1	2	7	5	4	8	9
9	6	3	4	2	7	1	5	8
4	8	5	3	1	9	6	2	7
2	1	7	5	6	8	9	4	3
1	4	8	7	5	2	3	9	6
7	5	6	9	3	4	8	1	2
3	9	2	6	8	1	5	7	4

37

6	7	4	5	2	1	9	8	3
8	1	3	4	9	7	2	5	6
2	9	5	6	8	3	1	7	4
3	5	2	8	6	9	4	1	7
9	8	1	3	7	4	5	6	2
4	6	7	2	1	5	8	3	9
5	4	8	9	3	6	7	2	1
7	3	9	1	5	2	6	4	8
1	2	6	7	4	8	3	9	5

38

6	8	9	3	4	1	7	5	2
5	7	3	9	2	6	4	1	8
2	4	1	7	8	5	6	9	3
1	5	2	4	6	7	8	3	9
3	9	4	1	5	8	2	7	6
7	6	8	2	3	9	1	4	5
4	1	5	6	9	2	3	8	7
8	2	7	5	1	3	9	6	4
9	3	6	8	7	4	5	2	1

39

6	7	1	4	5	3	2	9	8
4	8	3	7	2	9	1	5	6
5	9	2	8	6	1	4	3	7
3	6	5	2	1	7	8	4	9
8	2	9	3	4	6	5	7	1
1	4	7	9	8	5	6	2	3
9	1	8	5	3	4	7	6	2
2	3	4	6	7	8	9	1	5
7	5	6	1	9	2	3	8	4

40

2	4	3	9	7	5	8	6	1
1	9	5	2	6	8	7	3	4
6	8	7	3	4	1	9	2	5
7	1	2	4	9	6	5	8	3
9	3	4	5	8	7	2	1	6
5	6	8	1	3	2	4	7	9
3	2	9	8	1	4	6	5	7
4	5	6	7	2	3	1	9	8
8	7	1	6	5	9	3	4	2

41

6	5	2	7	4	1	8	9	3
1	3	9	6	5	8	2	4	7
7	8	4	2	9	3	5	6	1
9	1	5	3	6	4	7	2	8
3	4	6	8	7	2	9	1	5
8	2	7	9	1	5	6	3	4
5	9	8	1	3	6	4	7	2
4	7	1	5	2	9	3	8	6
2	6	3	4	8	7	1	5	9

42

4	8	6	9	2	7	3	5	1
9	5	2	1	4	3	8	7	6
1	7	3	8	6	5	2	4	9
8	6	4	2	5	9	7	1	3
3	2	1	6	7	4	9	8	5
7	9	5	3	8	1	4	6	2
6	4	8	5	9	2	1	3	7
2	1	7	4	3	6	5	9	8
5	3	9	7	1	8	6	2	4

43

5	1	8	4	6	3	7	9	2
6	2	4	5	7	9	3	1	8
9	3	7	1	8	2	4	6	5
7	6	9	2	4	5	1	8	3
1	8	2	7	3	6	9	5	4
3	4	5	9	1	8	6	2	7
8	7	1	6	2	4	5	3	9
4	9	3	8	5	1	2	7	6
2	5	6	3	9	7	8	4	1

44

8	6	5	2	1	9	7	4	3
2	9	7	4	8	3	1	5	6
3	1	4	7	6	5	9	8	2
4	3	1	9	5	6	8	2	7
9	7	2	8	4	1	6	3	5
6	5	8	3	7	2	4	9	1
7	8	3	6	2	4	5	1	9
1	4	9	5	3	7	2	6	8
5	2	6	1	9	8	3	7	4

45

6	1	5	8	7	9	3	4	2
9	3	2	5	4	6	8	1	7
8	4	7	1	2	3	6	9	5
7	5	1	4	6	8	9	2	3
3	8	4	9	5	2	7	6	1
2	6	9	7	3	1	4	5	8
5	7	8	2	9	4	1	3	6
1	9	6	3	8	5	2	7	4
4	2	3	6	1	7	5	8	9

46

7	1	3	2	4	6	9	5	8
5	6	9	8	7	3	1	2	4
4	8	2	5	9	1	3	7	6
8	4	1	3	6	2	5	9	7
2	7	6	1	5	9	8	4	3
9	3	5	7	8	4	2	6	1
3	9	4	6	2	8	7	1	5
6	5	8	9	1	7	4	3	2
1	2	7	4	3	5	6	8	9

47

5	1	7	4	8	3	2	6	9
6	2	3	5	9	1	8	4	7
8	9	4	6	7	2	3	1	5
1	5	9	7	6	8	4	2	3
7	8	6	3	2	4	5	9	1
4	3	2	9	1	5	7	8	6
2	6	8	1	5	7	9	3	4
9	4	5	8	3	6	1	7	2
3	7	1	2	4	9	6	5	8

48

7	3	6	4	5	1	8	2	9
2	1	9	6	8	3	4	5	7
5	8	4	9	7	2	6	3	1
4	2	3	5	6	7	9	1	8
9	6	5	8	1	4	2	7	3
1	7	8	2	3	9	5	4	6
8	4	1	7	2	6	3	9	5
6	9	7	3	4	5	1	8	2
3	5	2	1	9	8	7	6	4

49

4	7	1	5	3	6	2	8	9
6	8	9	2	4	7	3	1	5
2	3	5	1	9	8	6	4	7
5	4	7	8	1	3	9	6	2
9	6	2	7	5	4	1	3	8
3	1	8	9	6	2	7	5	4
1	5	6	4	2	9	8	7	3
8	9	3	6	7	5	4	2	1
7	2	4	3	8	1	5	9	6

50

2	7	1	8	5	6	4	9	3
5	6	8	4	3	9	2	7	1
9	4	3	2	1	7	6	8	5
6	2	5	1	8	3	9	4	7
3	8	9	5	7	4	1	2	6
7	1	4	6	9	2	5	3	8
8	9	2	3	6	1	7	5	4
1	5	7	9	4	8	3	6	2
4	3	6	7	2	5	8	1	9

51

9	2	5	1	7	8	4	6	3
3	4	8	2	9	6	1	5	7
6	1	7	5	4	3	8	2	9
5	7	1	4	2	9	3	8	6
8	3	2	6	5	1	9	7	4
4	9	6	3	8	7	2	1	5
2	8	3	9	6	5	7	4	1
1	6	4	7	3	2	5	9	8
7	5	9	8	1	4	6	3	2

52

4	6	9	7	1	3	2	5	8
3	2	5	8	9	6	1	7	4
1	8	7	4	5	2	6	3	9
5	1	2	3	4	8	9	6	7
7	3	4	5	6	9	8	1	2
6	9	8	2	7	1	3	4	5
8	7	3	1	2	4	5	9	6
9	4	1	6	8	5	7	2	3
2	5	6	9	3	7	4	8	1

53

9	2	8	5	7	6	1	4	3
6	1	5	3	8	4	2	7	9
4	3	7	9	2	1	8	5	6
7	5	4	1	3	8	9	6	2
3	6	1	7	9	2	5	8	4
2	8	9	4	6	5	3	1	7
8	9	6	2	5	7	4	3	1
1	7	2	8	4	3	6	9	5
5	4	3	6	1	9	7	2	8

54

1	8	4	9	6	2	3	7	5
6	2	5	4	7	3	8	1	9
7	9	3	5	8	1	4	6	2
2	4	7	1	3	5	9	8	6
9	1	6	8	2	7	5	4	3
5	3	8	6	9	4	1	2	7
4	6	9	2	5	8	7	3	1
8	7	2	3	1	9	6	5	4
3	5	1	7	4	6	2	9	8

55

5	3	2	7	8	1	9	6	4
1	7	8	4	9	6	3	5	2
6	4	9	5	2	3	8	1	7
9	6	7	8	1	2	4	3	5
8	5	1	9	3	4	2	7	6
3	2	4	6	5	7	1	9	8
2	1	6	3	7	8	5	4	9
4	8	5	1	6	9	7	2	3
7	9	3	2	4	5	6	8	1

56

1	5	2	3	9	4	8	6	7
3	6	9	5	7	8	1	4	2
8	4	7	1	6	2	9	3	5
9	8	4	6	1	5	7	2	3
6	3	1	4	2	7	5	9	8
2	7	5	9	8	3	6	1	4
7	1	8	2	3	9	4	5	6
4	2	6	8	5	1	3	7	9
5	9	3	7	4	6	2	8	1

57

6	1	8	3	9	4	7	5	2
5	4	3	6	2	7	1	9	8
7	9	2	8	5	1	3	6	4
4	6	7	2	1	3	9	8	5
2	8	5	7	4	9	6	1	3
1	3	9	5	8	6	4	2	7
9	5	1	4	7	8	2	3	6
3	2	4	9	6	5	8	7	1
8	7	6	1	3	2	5	4	9

58

3	9	8	7	6	5	2	4	1
1	4	7	9	2	8	5	6	3
5	2	6	4	1	3	9	8	7
2	6	5	8	7	1	3	9	4
4	3	9	6	5	2	1	7	8
8	7	1	3	4	9	6	5	2
6	8	2	5	3	7	4	1	9
7	5	3	1	9	4	8	2	6
9	1	4	2	8	6	7	3	5

59

5	4	6	3	1	9	7	8	2
8	1	3	2	5	7	4	9	6
7	2	9	6	4	8	1	5	3
9	6	1	5	8	3	2	4	7
4	7	5	1	6	2	8	3	9
2	3	8	9	7	4	5	6	1
3	5	7	4	2	6	9	1	8
6	8	4	7	9	1	3	2	5
1	9	2	8	3	5	6	7	4

60

2	8	4	5	6	9	1	3	7
9	7	6	3	2	1	4	5	8
1	3	5	4	7	8	6	9	2
7	4	2	8	5	6	3	1	9
3	1	8	2	9	4	7	6	5
6	5	9	1	3	7	2	8	4
8	9	3	7	1	2	5	4	6
5	6	7	9	4	3	8	2	1
4	2	1	6	8	5	9	7	3

61

8	5	3	2	6	4	7	9	1
2	4	6	9	7	1	3	5	8
7	1	9	3	5	8	4	6	2
4	8	1	7	2	6	9	3	5
9	3	2	1	4	5	6	8	7
6	7	5	8	9	3	2	1	4
3	6	7	5	8	2	1	4	9
5	9	4	6	1	7	8	2	3
1	2	8	4	3	9	5	7	6

62

6	9	4	2	1	8	7	5	3
7	2	8	9	5	3	1	4	6
3	5	1	7	6	4	2	9	8
9	4	5	6	2	1	3	8	7
2	1	7	3	8	9	5	6	4
8	6	3	5	4	7	9	1	2
4	8	2	1	7	5	6	3	9
5	3	6	8	9	2	4	7	1
1	7	9	4	3	6	8	2	5

63

3	5	9	8	1	2	4	7	6
1	8	6	9	7	4	5	2	3
7	4	2	5	3	6	1	8	9
2	3	4	1	9	5	7	6	8
8	1	5	7	6	3	9	4	2
6	9	7	4	2	8	3	1	5
4	2	3	6	5	1	8	9	7
9	6	1	3	8	7	2	5	4
5	7	8	2	4	9	6	3	1

64

2	1	3	6	5	9	7	4	8
5	4	6	2	7	8	1	9	3
8	7	9	4	1	3	5	6	2
3	8	5	7	9	6	4	2	1
6	9	4	5	2	1	8	3	7
1	2	7	3	8	4	6	5	9
4	3	8	1	6	2	9	7	5
7	6	1	9	3	5	2	8	4
9	5	2	8	4	7	3	1	6

65

8	3	9	2	4	6	5	7	1
4	1	2	5	7	9	3	8	6
5	6	7	3	8	1	4	2	9
3	4	6	9	2	7	8	1	5
2	5	1	4	6	8	7	9	3
7	9	8	1	3	5	6	4	2
1	8	4	6	5	2	9	3	7
6	2	3	7	9	4	1	5	8
9	7	5	8	1	3	2	6	4

66

4	9	2	6	8	1	3	7	5
7	8	5	3	4	9	2	6	1
1	6	3	2	5	7	9	8	4
2	5	4	7	9	3	6	1	8
9	1	8	4	2	6	5	3	7
3	7	6	5	1	8	4	9	2
6	2	7	1	3	4	8	5	9
8	4	1	9	6	5	7	2	3
5	3	9	8	7	2	1	4	6

67

6	1	5	3	2	4	9	7	8
3	9	2	5	8	7	4	1	6
7	4	8	9	1	6	3	2	5
4	2	1	8	9	5	6	3	7
9	8	3	6	7	1	5	4	2
5	6	7	4	3	2	1	8	9
2	7	6	1	4	9	8	5	3
1	3	9	2	5	8	7	6	4
8	5	4	7	6	3	2	9	1

68

2	8	6	7	1	5	4	9	3
9	4	1	2	8	3	5	7	6
3	7	5	4	6	9	8	1	2
6	2	9	8	3	4	1	5	7
4	1	8	6	5	7	2	3	9
5	3	7	9	2	1	6	4	8
7	9	2	5	4	8	3	6	1
1	6	4	3	9	2	7	8	5
8	5	3	1	7	6	9	2	4

69

8	3	7	9	5	2	4	6	1
6	9	5	4	1	3	8	2	7
1	4	2	8	7	6	5	3	9
7	8	6	1	9	5	3	4	2
9	5	3	2	6	4	1	7	8
4	2	1	3	8	7	9	5	6
5	6	8	7	3	1	2	9	4
2	7	9	5	4	8	6	1	3
3	1	4	6	2	9	7	8	5

70

8	5	4	3	6	2	1	7	9
9	1	7	5	8	4	2	6	3
3	2	6	9	7	1	5	8	4
6	7	2	4	9	8	3	1	5
4	8	3	2	1	5	7	9	6
5	9	1	7	3	6	8	4	2
7	6	5	1	2	9	4	3	8
1	4	8	6	5	3	9	2	7
2	3	9	8	4	7	6	5	1

71

6	3	2	4	5	1	7	8	9
9	5	1	6	7	8	3	4	2
8	4	7	2	3	9	6	5	1
2	1	5	8	4	3	9	6	7
4	8	9	1	6	7	2	3	5
3	7	6	5	9	2	4	1	8
7	9	8	3	1	4	5	2	6
1	6	4	7	2	5	8	9	3
5	2	3	9	8	6	1	7	4

72

4	1	2	9	6	3	7	5	8
8	5	6	1	2	7	4	9	3
3	7	9	8	5	4	1	6	2
5	3	8	6	1	9	2	7	4
1	9	7	2	4	5	8	3	6
2	6	4	3	7	8	5	1	9
9	8	5	4	3	1	6	2	7
6	4	1	7	9	2	3	8	5
7	2	3	5	8	6	9	4	1

73

2	1	8	3	5	7	6	4	9
7	5	9	1	6	4	3	8	2
3	6	4	8	9	2	5	7	1
6	9	7	4	2	5	8	1	3
4	8	1	9	7	3	2	5	6
5	2	3	6	1	8	7	9	4
1	3	5	7	4	6	9	2	8
9	7	6	2	8	1	4	3	5
8	4	2	5	3	9	1	6	7

74

5	6	7	4	8	2	1	3	9
4	2	3	1	5	9	7	8	6
9	8	1	3	7	6	2	4	5
3	1	2	5	6	7	8	9	4
8	5	6	2	9	4	3	1	7
7	4	9	8	1	3	6	5	2
6	3	5	9	2	1	4	7	8
1	7	8	6	4	5	9	2	3
2	9	4	7	3	8	5	6	1

75

6	7	4	1	2	9	5	8	3
3	1	5	8	7	6	2	4	9
8	2	9	4	5	3	1	7	6
2	8	1	3	9	5	4	6	7
5	9	7	2	6	4	3	1	8
4	3	6	7	8	1	9	5	2
1	4	8	6	3	2	7	9	5
9	6	2	5	1	7	8	3	4
7	5	3	9	4	8	6	2	1

76

9	8	3	1	7	4	5	6	2
4	7	2	6	5	3	1	8	9
1	6	5	2	9	8	7	3	4
5	9	1	3	2	7	8	4	6
8	4	7	9	6	1	3	2	5
2	3	6	8	4	5	9	1	7
3	2	9	5	8	6	4	7	1
6	1	4	7	3	9	2	5	8
7	5	8	4	1	2	6	9	3

77

5	4	7	8	2	1	3	9	6
6	1	2	5	9	3	8	7	4
8	9	3	6	7	4	5	2	1
9	5	6	2	8	7	1	4	3
7	3	8	1	4	9	6	5	2
1	2	4	3	5	6	9	8	7
4	6	9	7	1	5	2	3	8
2	7	1	9	3	8	4	6	5
3	8	5	4	6	2	7	1	9

78

2	8	4	6	5	7	1	3	9
6	3	5	2	9	1	4	7	8
1	9	7	8	3	4	6	5	2
9	7	1	3	6	8	5	2	4
4	6	2	9	7	5	3	8	1
8	5	3	4	1	2	7	9	6
7	1	8	5	2	6	9	4	3
3	2	6	7	4	9	8	1	5
5	4	9	1	8	3	2	6	7

79

1	4	5	9	8	3	7	6	2
6	2	7	1	4	5	8	9	3
9	8	3	2	6	7	1	5	4
5	9	2	4	1	6	3	7	8
4	1	8	7	3	9	5	2	6
3	7	6	5	2	8	9	4	1
2	5	1	8	7	4	6	3	9
7	3	4	6	9	1	2	8	5
8	6	9	3	5	2	4	1	7

80

7	1	8	3	6	9	4	2	5
2	6	5	4	8	7	3	1	9
4	9	3	5	2	1	6	7	8
1	7	6	2	9	8	5	4	3
9	8	4	7	5	3	2	6	1
3	5	2	1	4	6	9	8	7
5	4	7	8	3	2	1	9	6
6	2	1	9	7	5	8	3	4
8	3	9	6	1	4	7	5	2

81

6	8	5	4	9	1	2	7	3
2	7	1	8	6	3	5	4	9
3	9	4	7	5	2	8	6	1
4	5	6	3	2	9	1	8	7
9	2	7	6	1	8	4	3	5
1	3	8	5	7	4	9	2	6
7	1	3	2	4	5	6	9	8
8	4	9	1	3	6	7	5	2
5	6	2	9	8	7	3	1	4

82

7	2	1	3	9	4	8	6	5
6	3	5	7	1	8	9	4	2
9	8	4	5	2	6	1	7	3
8	6	9	4	3	7	5	2	1
3	5	2	6	8	1	4	9	7
1	4	7	2	5	9	3	8	6
2	1	6	9	4	3	7	5	8
4	7	8	1	6	5	2	3	9
5	9	3	8	7	2	6	1	4

83

9	3	5	1	7	6	8	2	4
2	6	8	4	9	5	3	1	7
1	7	4	2	3	8	6	9	5
4	8	6	9	5	7	2	3	1
5	1	7	6	2	3	4	8	9
3	2	9	8	4	1	7	5	6
8	9	2	7	1	4	5	6	3
6	4	3	5	8	9	1	7	2
7	5	1	3	6	2	9	4	8

84

6	8	5	4	1	7	2	3	9
7	2	1	3	5	9	4	8	6
3	9	4	2	8	6	5	1	7
4	3	8	9	7	2	1	6	5
2	6	7	1	3	5	9	4	8
5	1	9	8	6	4	7	2	3
9	5	2	6	4	8	3	7	1
8	4	3	7	9	1	6	5	2
1	7	6	5	2	3	8	9	4

85

5	7	6	8	3	1	4	9	2
4	1	2	5	9	6	7	8	3
9	8	3	4	7	2	6	1	5
7	3	1	9	2	8	5	4	6
2	5	9	1	6	4	8	3	7
6	4	8	3	5	7	9	2	1
1	6	5	2	4	9	3	7	8
8	9	7	6	1	3	2	5	4
3	2	4	7	8	5	1	6	9

86

8	7	5	2	6	4	1	3	9
4	6	3	7	1	9	8	2	5
2	9	1	3	5	8	6	4	7
3	1	2	5	8	7	4	9	6
6	4	8	9	3	1	5	7	2
9	5	7	4	2	6	3	8	1
5	8	4	1	9	2	7	6	3
7	3	9	6	4	5	2	1	8
1	2	6	8	7	3	9	5	4

87

2	5	6	9	4	3	8	7	1
3	9	8	2	1	7	6	5	4
4	1	7	6	5	8	9	3	2
8	3	9	1	2	5	7	4	6
5	7	2	4	8	6	3	1	9
6	4	1	7	3	9	2	8	5
9	8	5	3	6	1	4	2	7
1	6	4	8	7	2	5	9	3
7	2	3	5	9	4	1	6	8

88

4	2	5	9	3	1	7	8	6
7	1	3	8	5	6	9	4	2
6	9	8	4	7	2	1	3	5
8	5	6	1	9	3	2	7	4
2	7	4	6	8	5	3	9	1
1	3	9	2	4	7	5	6	8
5	6	7	3	2	8	4	1	9
9	8	2	7	1	4	6	5	3
3	4	1	5	6	9	8	2	7

89

7	5	6	8	4	1	3	9	2
9	1	4	6	3	2	8	7	5
3	8	2	5	9	7	1	4	6
5	7	8	1	6	9	2	3	4
2	4	9	3	5	8	6	1	7
6	3	1	2	7	4	9	5	8
4	9	3	7	2	6	5	8	1
8	6	7	9	1	5	4	2	3
1	2	5	4	8	3	7	6	9

90

4	8	1	9	5	7	6	2	3
5	2	9	4	6	3	7	8	1
7	3	6	1	2	8	5	4	9
2	6	5	8	9	4	1	3	7
9	1	3	6	7	2	4	5	8
8	4	7	3	1	5	2	9	6
1	7	4	2	3	9	8	6	5
3	5	2	7	8	6	9	1	4
6	9	8	5	4	1	3	7	2

91

8	6	2	3	5	4	7	9	1
4	1	7	8	2	9	5	3	6
9	3	5	7	6	1	2	4	8
7	8	6	1	3	2	4	5	9
5	4	1	6	9	7	8	2	3
2	9	3	5	4	8	6	1	7
6	5	8	2	1	3	9	7	4
1	2	9	4	7	6	3	8	5
3	7	4	9	8	5	1	6	2

92

6	9	4	3	8	7	2	1	5
7	3	2	4	5	1	9	8	6
8	1	5	9	2	6	3	4	7
4	2	7	1	6	5	8	3	9
1	5	6	8	9	3	7	2	4
9	8	3	2	7	4	6	5	1
2	6	1	7	4	8	5	9	3
3	7	8	5	1	9	4	6	2
5	4	9	6	3	2	1	7	8

93

3	7	8	9	5	1	4	6	2
5	2	6	4	8	3	7	9	1
9	4	1	7	6	2	8	3	5
6	3	9	1	2	7	5	4	8
7	8	4	5	3	6	1	2	9
2	1	5	8	9	4	6	7	3
4	6	3	2	1	8	9	5	7
8	9	7	3	4	5	2	1	6
1	5	2	6	7	9	3	8	4

94

1	9	5	7	8	3	6	2	4
7	4	3	5	6	2	8	9	1
2	6	8	1	9	4	5	7	3
5	2	4	8	7	6	1	3	9
8	1	9	2	3	5	4	6	7
6	3	7	4	1	9	2	5	8
9	5	2	3	4	1	7	8	6
3	8	1	6	5	7	9	4	2
4	7	6	9	2	8	3	1	5

95

2	4	7	5	6	9	3	8	1
1	3	6	8	2	7	9	4	5
8	9	5	1	4	3	2	6	7
4	1	3	9	8	2	5	7	6
6	2	9	4	7	5	1	3	8
5	7	8	3	1	6	4	9	2
3	5	1	6	9	8	7	2	4
7	8	4	2	3	1	6	5	9
9	6	2	7	5	4	8	1	3

96

2	6	9	1	3	4	5	8	7
8	3	7	6	5	9	1	2	4
5	1	4	2	8	7	9	3	6
6	5	3	8	2	1	7	4	9
4	2	8	7	9	6	3	5	1
7	9	1	5	4	3	8	6	2
9	4	2	3	7	5	6	1	8
1	7	5	4	6	8	2	9	3
3	8	6	9	1	2	4	7	5

97

4	1	6	2	5	7	9	3	8
2	7	8	6	9	3	4	5	1
5	9	3	4	8	1	7	6	2
6	4	5	1	2	9	8	7	3
7	3	2	8	4	6	5	1	9
1	8	9	7	3	5	6	2	4
8	6	1	9	7	2	3	4	5
9	5	7	3	1	4	2	8	6
3	2	4	5	6	8	1	9	7

98

4	5	6	7	1	3	9	8	2
8	1	2	6	9	4	5	7	3
9	7	3	2	5	8	1	6	4
7	2	8	5	3	9	6	4	1
3	6	9	1	4	7	8	2	5
1	4	5	8	2	6	3	9	7
2	9	1	4	8	5	7	3	6
6	8	4	3	7	1	2	5	9
5	3	7	9	6	2	4	1	8

99

5	1	2	8	9	7	4	6	3
3	6	7	2	1	4	8	9	5
8	4	9	6	3	5	2	1	7
6	8	1	5	2	9	7	3	4
4	9	3	7	8	1	6	5	2
2	7	5	3	4	6	1	8	9
9	5	6	1	7	2	3	4	8
1	2	8	4	5	3	9	7	6
7	3	4	9	6	8	5	2	1

100

7	8	9	4	6	3	2	1	5
2	4	6	7	5	1	8	3	9
1	3	5	9	2	8	4	7	6
5	1	3	6	7	2	9	4	8
6	2	7	8	9	4	1	5	3
4	9	8	1	3	5	7	6	2
8	7	2	3	1	6	5	9	4
3	5	1	2	4	9	6	8	7
9	6	4	5	8	7	3	2	1

101

2	6	9	4	1	8	5	7	3
5	4	7	3	2	6	1	9	8
3	1	8	7	9	5	2	4	6
6	7	5	8	3	4	9	1	2
9	2	3	6	5	1	7	8	4
4	8	1	2	7	9	3	6	5
8	3	2	1	6	7	4	5	9
1	9	6	5	4	3	8	2	7
7	5	4	9	8	2	6	3	1

102

7	2	4	3	5	6	9	1	8
1	5	8	2	7	9	4	6	3
6	9	3	1	8	4	5	2	7
2	6	5	7	9	8	1	3	4
4	3	9	5	1	2	7	8	6
8	1	7	4	6	3	2	9	5
5	8	2	9	3	7	6	4	1
3	4	1	6	2	5	8	7	9
9	7	6	8	4	1	3	5	2

103

7	2	5	1	9	8	3	6	4
4	1	9	6	5	3	2	7	8
6	8	3	7	4	2	9	1	5
9	6	2	8	3	1	4	5	7
8	5	1	4	2	7	6	3	9
3	7	4	9	6	5	8	2	1
1	4	6	3	7	9	5	8	2
5	3	8	2	1	4	7	9	6
2	9	7	5	8	6	1	4	3

104

9	2	8	7	1	3	4	5	6
4	1	5	6	9	2	8	7	3
7	6	3	5	4	8	9	2	1
5	4	9	3	2	6	7	1	8
1	3	7	9	8	5	6	4	2
6	8	2	4	7	1	3	9	5
8	9	1	2	6	4	5	3	7
3	7	6	1	5	9	2	8	4
2	5	4	8	3	7	1	6	9

105

1	5	3	2	7	4	9	6	8
2	9	4	3	6	8	1	7	5
8	6	7	1	9	5	3	4	2
3	4	8	5	2	6	7	9	1
7	2	5	4	1	9	8	3	6
6	1	9	8	3	7	2	5	4
9	3	2	6	4	1	5	8	7
5	7	6	9	8	2	4	1	3
4	8	1	7	5	3	6	2	9

106

1	2	5	4	8	7	9	3	6
3	4	9	2	5	6	7	1	8
8	7	6	9	3	1	4	2	5
5	9	8	6	4	2	1	7	3
7	1	4	5	9	3	8	6	2
2	6	3	1	7	8	5	4	9
4	3	7	8	6	5	2	9	1
6	5	1	7	2	9	3	8	4
9	8	2	3	1	4	6	5	7

107

7	5	6	3	8	2	1	4	9
8	4	9	6	7	1	3	2	5
2	1	3	4	9	5	7	8	6
4	8	7	1	3	9	6	5	2
6	2	1	8	5	7	9	3	4
9	3	5	2	4	6	8	1	7
3	7	4	9	2	8	5	6	1
5	6	2	7	1	3	4	9	8
1	9	8	5	6	4	2	7	3

108

8	3	1	6	5	7	2	9	4
7	2	4	9	3	1	6	5	8
9	6	5	2	8	4	7	3	1
6	1	2	5	7	9	8	4	3
4	5	8	3	6	2	1	7	9
3	9	7	4	1	8	5	2	6
5	8	6	7	9	3	4	1	2
1	4	3	8	2	5	9	6	7
2	7	9	1	4	6	3	8	5

109

6	2	4	1	9	3	5	8	7
7	8	5	6	2	4	3	9	1
1	9	3	8	5	7	2	4	6
9	4	2	7	8	6	1	5	3
8	1	6	3	4	5	9	7	2
3	5	7	9	1	2	8	6	4
4	6	9	2	3	8	7	1	5
5	3	8	4	7	1	6	2	9
2	7	1	5	6	9	4	3	8

110

7	8	1	2	5	6	9	4	3
2	3	5	7	9	4	8	1	6
9	6	4	1	8	3	7	2	5
3	7	6	4	1	5	2	8	9
4	9	8	3	2	7	5	6	1
5	1	2	9	6	8	4	3	7
6	5	3	8	4	9	1	7	2
8	2	7	5	3	1	6	9	4
1	4	9	6	7	2	3	5	8

111

9	4	8	2	6	7	1	3	5
3	6	7	4	5	1	8	9	2
5	2	1	8	3	9	4	7	6
7	3	2	9	8	6	5	4	1
4	8	5	7	1	2	9	6	3
1	9	6	5	4	3	2	8	7
6	5	4	3	2	8	7	1	9
8	1	9	6	7	5	3	2	4
2	7	3	1	9	4	6	5	8

112

9	6	5	8	1	4	2	3	7
4	3	1	9	2	7	6	8	5
2	8	7	5	6	3	9	1	4
3	1	8	7	9	2	4	5	6
5	7	4	1	3	6	8	9	2
6	2	9	4	8	5	3	7	1
1	4	6	3	7	9	5	2	8
8	9	2	6	5	1	7	4	3
7	5	3	2	4	8	1	6	9

113

5	6	7	3	8	9	4	1	2
2	1	4	7	5	6	8	3	9
9	8	3	2	4	1	7	6	5
7	3	6	5	1	4	9	2	8
4	2	9	8	6	7	1	5	3
1	5	8	9	2	3	6	7	4
3	7	1	4	9	2	5	8	6
8	4	2	6	7	5	3	9	1
6	9	5	1	3	8	2	4	7

114

4	9	6	3	8	2	1	7	5
2	3	5	9	7	1	8	6	4
1	8	7	6	5	4	3	2	9
3	2	1	7	9	5	4	8	6
7	6	4	1	2	8	5	9	3
9	5	8	4	3	6	2	1	7
8	7	2	5	6	3	9	4	1
6	4	3	8	1	9	7	5	2
5	1	9	2	4	7	6	3	8

115

8	4	5	3	6	7	9	1	2
1	3	2	4	9	8	5	6	7
9	7	6	5	1	2	4	8	3
6	8	7	1	3	4	2	5	9
4	5	3	6	2	9	8	7	1
2	1	9	7	8	5	3	4	6
7	6	8	9	4	3	1	2	5
3	2	1	8	5	6	7	9	4
5	9	4	2	7	1	6	3	8

116

3	8	5	1	2	9	6	7	4
2	7	6	5	4	3	1	8	9
9	4	1	8	6	7	2	5	3
6	3	4	7	8	1	5	9	2
7	2	8	6	9	5	4	3	1
5	1	9	4	3	2	8	6	7
4	6	7	3	1	8	9	2	5
8	5	2	9	7	4	3	1	6
1	9	3	2	5	6	7	4	8

117

5	4	3	6	8	1	2	7	9
7	1	8	9	5	2	6	4	3
9	6	2	4	3	7	5	8	1
4	8	1	7	6	9	3	2	5
3	5	6	8	2	4	1	9	7
2	7	9	5	1	3	8	6	4
6	3	7	2	9	5	4	1	8
1	2	4	3	7	8	9	5	6
8	9	5	1	4	6	7	3	2

118

1	9	5	8	3	6	2	7	4
3	4	6	5	7	2	1	8	9
8	2	7	1	4	9	6	3	5
9	6	4	3	8	5	7	1	2
2	3	8	9	1	7	4	5	6
7	5	1	2	6	4	3	9	8
4	8	9	7	2	1	5	6	3
6	1	3	4	5	8	9	2	7
5	7	2	6	9	3	8	4	1

119

8	2	5	1	3	6	9	4	7
3	7	6	9	5	4	8	2	1
4	9	1	8	7	2	6	3	5
1	3	7	4	8	5	2	9	6
6	5	2	7	9	3	1	8	4
9	8	4	6	2	1	7	5	3
5	4	9	2	6	7	3	1	8
7	1	8	3	4	9	5	6	2
2	6	3	5	1	8	4	7	9

120

9	5	4	3	7	6	8	1	2
3	2	7	1	5	8	6	4	9
6	1	8	2	9	4	5	7	3
8	4	5	7	1	9	2	3	6
7	9	6	8	3	2	1	5	4
1	3	2	4	6	5	9	8	7
5	8	9	6	4	3	7	2	1
4	6	1	5	2	7	3	9	8
2	7	3	9	8	1	4	6	5

121

2	4	6	9	8	7	5	1	3
7	9	3	6	5	1	2	4	8
8	5	1	3	2	4	7	6	9
9	8	4	2	6	5	3	7	1
3	1	5	7	4	8	6	9	2
6	2	7	1	3	9	8	5	4
5	3	9	4	7	2	1	8	6
4	7	2	8	1	6	9	3	5
1	6	8	5	9	3	4	2	7

122

5	9	7	2	4	3	1	8	6
2	3	1	9	8	6	7	5	4
6	4	8	1	7	5	3	2	9
4	8	2	7	6	9	5	1	3
9	1	3	5	2	8	4	6	7
7	6	5	3	1	4	8	9	2
1	7	9	4	5	2	6	3	8
3	5	6	8	9	7	2	4	1
8	2	4	6	3	1	9	7	5

123

5	3	2	6	8	4	9	1	7
9	4	7	1	3	2	6	5	8
6	8	1	5	9	7	2	3	4
2	6	8	9	1	5	7	4	3
1	9	4	7	6	3	8	2	5
7	5	3	2	4	8	1	6	9
8	2	6	3	5	9	4	7	1
3	1	9	4	7	6	5	8	2
4	7	5	8	2	1	3	9	6

124

8	5	1	2	9	4	7	3	6
4	9	3	6	7	1	2	8	5
7	2	6	8	5	3	9	1	4
2	6	4	1	3	5	8	7	9
1	3	9	4	8	7	5	6	2
5	7	8	9	2	6	1	4	3
9	1	5	3	4	8	6	2	7
3	8	2	7	6	9	4	5	1
6	4	7	5	1	2	3	9	8

125

1	3	9	7	4	2	8	5	6
2	8	4	5	9	6	3	7	1
5	6	7	8	3	1	9	4	2
7	5	3	2	6	8	1	9	4
6	1	2	9	7	4	5	8	3
4	9	8	3	1	5	2	6	7
8	4	6	1	5	3	7	2	9
3	7	5	4	2	9	6	1	8
9	2	1	6	8	7	4	3	5

126

3	9	6	8	1	7	2	4	5
7	1	4	3	5	2	8	9	6
2	8	5	6	9	4	3	1	7
4	7	8	5	2	3	1	6	9
9	5	2	1	6	8	4	7	3
6	3	1	7	4	9	5	2	8
8	6	7	4	3	1	9	5	2
1	2	3	9	7	5	6	8	4
5	4	9	2	8	6	7	3	1

127

5	6	8	2	9	4	1	3	7
9	3	7	8	5	1	6	4	2
4	2	1	3	6	7	8	9	5
3	8	6	1	4	5	7	2	9
7	9	5	6	8	2	4	1	3
1	4	2	7	3	9	5	6	8
2	7	9	5	1	6	3	8	4
6	5	3	4	2	8	9	7	1
8	1	4	9	7	3	2	5	6

128

9	2	5	7	6	1	8	4	3
4	8	7	2	5	3	1	9	6
6	1	3	4	8	9	2	5	7
3	4	1	9	7	8	6	2	5
5	9	6	3	4	2	7	8	1
2	7	8	6	1	5	9	3	4
1	5	9	8	3	7	4	6	2
8	3	4	1	2	6	5	7	9
7	6	2	5	9	4	3	1	8

129

4	8	9	7	2	5	1	6	3
6	5	3	1	4	9	7	8	2
2	1	7	6	8	3	4	5	9
9	2	1	3	6	4	5	7	8
5	3	4	2	7	8	6	9	1
8	7	6	9	5	1	2	3	4
1	6	5	8	3	2	9	4	7
3	4	2	5	9	7	8	1	6
7	9	8	4	1	6	3	2	5

130

6	1	3	4	8	7	2	5	9
9	8	4	3	2	5	7	6	1
5	7	2	6	9	1	4	3	8
8	3	5	1	6	2	9	7	4
1	4	9	8	7	3	6	2	5
7	2	6	5	4	9	1	8	3
4	5	1	7	3	6	8	9	2
2	6	8	9	5	4	3	1	7
3	9	7	2	1	8	5	4	6

131

5	1	2	3	6	9	7	4	8
7	3	8	1	4	2	9	5	6
6	4	9	5	8	7	3	2	1
2	7	1	4	5	3	6	8	9
9	6	4	2	7	8	1	3	5
8	5	3	6	9	1	2	7	4
4	2	7	9	1	5	8	6	3
1	8	6	7	3	4	5	9	2
3	9	5	8	2	6	4	1	7

132

8	7	1	5	4	6	9	2	3
2	3	4	1	7	9	5	8	6
9	6	5	3	2	8	7	1	4
5	4	7	6	3	1	2	9	8
3	9	8	4	5	2	6	7	1
6	1	2	9	8	7	3	4	5
1	5	6	7	9	4	8	3	2
4	2	9	8	6	3	1	5	7
7	8	3	2	1	5	4	6	9

133

3	4	9	2	8	5	1	7	6
5	8	1	9	6	7	4	2	3
7	2	6	3	4	1	5	9	8
6	5	7	1	3	4	9	8	2
2	3	4	5	9	8	6	1	7
9	1	8	7	2	6	3	4	5
4	7	2	6	1	3	8	5	9
8	6	5	4	7	9	2	3	1
1	9	3	8	5	2	7	6	4

134

6	3	1	4	7	2	5	8	9
8	2	5	9	3	1	4	6	7
4	7	9	5	8	6	1	2	3
7	5	2	8	6	9	3	4	1
3	6	4	2	1	5	9	7	8
9	1	8	3	4	7	6	5	2
5	8	7	1	9	4	2	3	6
2	9	3	6	5	8	7	1	4
1	4	6	7	2	3	8	9	5

135

7	4	3	2	1	8	9	5	6
1	5	9	7	3	6	2	8	4
8	2	6	9	4	5	7	3	1
6	9	5	3	2	1	8	4	7
2	1	8	6	7	4	3	9	5
4	3	7	5	8	9	1	6	2
9	7	2	4	5	3	6	1	8
3	8	4	1	6	2	5	7	9
5	6	1	8	9	7	4	2	3

136

9	1	5	4	7	3	8	2	6
2	7	8	1	5	6	3	4	9
3	4	6	2	9	8	7	5	1
4	3	2	5	8	9	1	6	7
7	8	1	3	6	2	5	9	4
5	6	9	7	4	1	2	3	8
8	9	3	6	2	7	4	1	5
1	5	7	9	3	4	6	8	2
6	2	4	8	1	5	9	7	3

137

3	7	9	2	1	4	6	5	8
8	6	2	7	3	5	4	9	1
5	1	4	6	8	9	3	7	2
2	3	6	8	5	7	9	1	4
1	8	7	4	9	6	2	3	5
9	4	5	3	2	1	7	8	6
4	2	1	5	7	3	8	6	9
6	9	3	1	4	8	5	2	7
7	5	8	9	6	2	1	4	3

138

3	7	1	9	4	8	5	2	6
4	2	8	6	7	5	1	9	3
5	6	9	1	2	3	7	8	4
1	5	4	3	6	9	8	7	2
2	3	7	8	1	4	9	6	5
9	8	6	7	5	2	3	4	1
8	1	3	4	9	6	2	5	7
7	4	2	5	8	1	6	3	9
6	9	5	2	3	7	4	1	8

139

4	8	6	1	2	3	7	5	9
3	2	1	7	5	9	6	4	8
5	7	9	8	6	4	1	3	2
6	3	2	5	7	8	4	9	1
8	5	7	9	4	1	2	6	3
9	1	4	2	3	6	5	8	7
2	4	5	3	8	7	9	1	6
1	6	8	4	9	2	3	7	5
7	9	3	6	1	5	8	2	4

140

3	8	1	7	5	4	2	6	9
5	7	4	9	6	2	3	1	8
2	9	6	8	3	1	4	5	7
7	5	9	1	8	3	6	2	4
1	2	3	6	4	7	9	8	5
4	6	8	5	2	9	7	3	1
8	3	2	4	9	5	1	7	6
6	4	7	2	1	8	5	9	3
9	1	5	3	7	6	8	4	2

141

7	6	8	5	9	2	1	3	4
9	4	5	3	1	8	6	7	2
1	3	2	4	7	6	5	9	8
4	8	6	9	5	7	3	2	1
5	9	3	6	2	1	4	8	7
2	7	1	8	4	3	9	5	6
3	5	7	1	8	4	2	6	9
6	2	4	7	3	9	8	1	5
8	1	9	2	6	5	7	4	3

142

4	1	8	7	6	9	2	3	5
2	3	5	1	4	8	9	6	7
7	6	9	2	3	5	8	1	4
1	9	4	6	5	7	3	2	8
3	2	7	8	9	4	1	5	6
5	8	6	3	1	2	7	4	9
9	4	2	5	8	3	6	7	1
6	5	3	9	7	1	4	8	2
8	7	1	4	2	6	5	9	3

143

4	7	8	9	2	3	6	5	1
2	6	3	5	1	8	9	4	7
9	1	5	6	7	4	2	8	3
8	4	2	7	6	9	3	1	5
1	3	9	8	5	2	7	6	4
6	5	7	4	3	1	8	9	2
7	2	4	1	8	6	5	3	9
3	8	1	2	9	5	4	7	6
5	9	6	3	4	7	1	2	8

144

3	8	1	9	2	4	7	5	6
5	4	9	8	7	6	3	2	1
7	6	2	3	5	1	8	9	4
2	5	3	7	4	8	6	1	9
8	7	4	6	1	9	2	3	5
1	9	6	5	3	2	4	7	8
6	2	8	1	9	3	5	4	7
9	3	7	4	8	5	1	6	2
4	1	5	2	6	7	9	8	3

145

5	8	9	6	4	3	2	7	1
3	4	6	7	2	1	5	9	8
7	1	2	9	8	5	3	4	6
2	5	1	4	7	9	6	8	3
6	9	3	2	1	8	4	5	7
8	7	4	3	5	6	9	1	2
9	6	5	8	3	7	1	2	4
4	3	7	1	9	2	8	6	5
1	2	8	5	6	4	7	3	9

146

1	9	4	6	2	5	8	7	3
6	2	5	7	3	8	9	1	4
7	8	3	9	4	1	6	2	5
5	4	9	3	1	2	7	6	8
2	1	8	5	7	6	4	3	9
3	7	6	8	9	4	1	5	2
9	6	1	2	8	3	5	4	7
8	5	2	4	6	7	3	9	1
4	3	7	1	5	9	2	8	6

147

4	7	6	5	2	8	1	3	9
8	3	9	6	4	1	5	2	7
1	5	2	9	3	7	4	8	6
3	4	7	2	8	5	9	6	1
2	1	5	3	9	6	7	4	8
6	9	8	1	7	4	3	5	2
9	6	1	8	5	3	2	7	4
7	8	3	4	1	2	6	9	5
5	2	4	7	6	9	8	1	3

148

9	6	4	8	2	5	7	1	3
1	3	2	7	6	9	5	4	8
8	7	5	1	4	3	2	9	6
3	1	6	4	7	2	8	5	9
5	9	7	6	3	8	4	2	1
2	4	8	9	5	1	3	6	7
7	2	3	5	9	6	1	8	4
6	5	1	3	8	4	9	7	2
4	8	9	2	1	7	6	3	5

149

2	1	7	5	4	9	8	3	6
5	4	8	7	6	3	1	9	2
3	9	6	1	8	2	4	5	7
9	2	1	6	3	4	7	8	5
4	6	5	2	7	8	3	1	9
7	8	3	9	1	5	2	6	4
8	3	9	4	5	7	6	2	1
1	7	2	3	9	6	5	4	8
6	5	4	8	2	1	9	7	3

150

5	4	9	2	7	8	3	6	1
3	6	2	1	9	4	8	5	7
1	8	7	6	3	5	9	2	4
2	5	6	9	8	7	1	4	3
4	3	8	5	2	1	7	9	6
9	7	1	3	4	6	2	8	5
7	1	4	8	6	2	5	3	9
6	2	3	7	5	9	4	1	8
8	9	5	4	1	3	6	7	2

151

1	2	5	7	4	8	6	3	9
3	7	9	2	6	5	8	4	1
6	4	8	9	3	1	7	5	2
8	6	2	4	1	7	5	9	3
4	3	1	6	5	9	2	7	8
9	5	7	8	2	3	4	1	6
2	1	4	3	7	6	9	8	5
7	9	3	5	8	2	1	6	4
5	8	6	1	9	4	3	2	7

152

4	9	8	1	2	7	3	6	5
6	1	2	4	3	5	8	7	9
7	5	3	9	8	6	1	2	4
9	6	5	8	7	4	2	3	1
3	2	4	5	6	1	7	9	8
8	7	1	3	9	2	4	5	6
1	3	9	7	5	8	6	4	2
5	4	6	2	1	3	9	8	7
2	8	7	6	4	9	5	1	3

153

6	5	9	2	8	4	3	7	1
3	8	7	1	9	5	6	4	2
1	2	4	6	7	3	9	8	5
4	3	5	9	6	8	2	1	7
9	6	1	7	5	2	8	3	4
2	7	8	4	3	1	5	9	6
7	1	3	8	2	6	4	5	9
8	9	2	5	4	7	1	6	3
5	4	6	3	1	9	7	2	8

154

2	3	4	7	1	9	5	8	6
8	5	6	2	3	4	9	1	7
9	7	1	5	8	6	3	2	4
1	6	9	8	5	2	4	7	3
5	8	7	3	4	1	2	6	9
4	2	3	9	6	7	8	5	1
3	4	5	6	7	8	1	9	2
6	1	2	4	9	5	7	3	8
7	9	8	1	2	3	6	4	5

155

1	2	9	4	8	7	5	6	3
4	6	3	9	2	5	1	8	7
8	5	7	6	3	1	4	9	2
5	8	1	7	6	9	2	3	4
9	3	4	1	5	2	6	7	8
2	7	6	3	4	8	9	5	1
6	1	8	5	7	4	3	2	9
7	4	5	2	9	3	8	1	6
3	9	2	8	1	6	7	4	5

156

1	7	3	5	6	4	8	9	2
8	4	6	3	2	9	5	7	1
2	5	9	7	1	8	3	6	4
9	1	8	2	4	5	7	3	6
3	6	4	1	9	7	2	5	8
5	2	7	6	8	3	4	1	9
4	3	2	9	7	1	6	8	5
6	9	5	8	3	2	1	4	7
7	8	1	4	5	6	9	2	3

157

5	2	4	6	7	8	3	9	1
3	1	7	2	4	9	8	6	5
6	9	8	5	3	1	4	2	7
8	4	1	3	5	6	2	7	9
9	3	2	7	1	4	6	5	8
7	5	6	9	8	2	1	3	4
4	6	5	8	2	7	9	1	3
2	8	3	1	9	5	7	4	6
1	7	9	4	6	3	5	8	2

158

1	6	9	4	8	3	5	2	7
4	2	3	7	1	5	9	8	6
5	8	7	2	9	6	1	4	3
6	3	1	9	5	8	2	7	4
7	4	8	3	2	1	6	5	9
2	9	5	6	7	4	3	1	8
3	7	2	1	4	9	8	6	5
8	1	6	5	3	7	4	9	2
9	5	4	8	6	2	7	3	1

159

1	5	9	6	3	4	8	7	2
3	8	6	2	5	7	9	4	1
7	2	4	8	9	1	3	6	5
4	6	5	9	8	2	1	3	7
9	3	1	4	7	5	2	8	6
8	7	2	3	1	6	4	5	9
6	1	8	5	4	9	7	2	3
5	9	3	7	2	8	6	1	4
2	4	7	1	6	3	5	9	8

160

7	9	4	5	8	6	1	3	2
3	1	2	4	9	7	8	5	6
8	6	5	1	3	2	4	7	9
6	8	1	2	5	9	3	4	7
2	3	7	6	4	1	5	9	8
4	5	9	3	7	8	6	2	1
9	4	6	8	2	3	7	1	5
1	2	3	7	6	5	9	8	4
5	7	8	9	1	4	2	6	3

161

7	8	1	3	6	2	4	9	5
5	9	3	1	8	4	2	6	7
2	4	6	5	9	7	3	8	1
4	3	5	2	7	9	8	1	6
8	2	9	6	3	1	5	7	4
1	6	7	8	4	5	9	3	2
3	5	4	9	1	6	7	2	8
6	7	8	4	2	3	1	5	9
9	1	2	7	5	8	6	4	3

162

5	1	2	9	3	6	8	4	7
4	8	3	7	5	2	6	1	9
6	9	7	4	8	1	2	5	3
9	3	6	2	4	5	1	7	8
8	5	4	1	7	9	3	6	2
7	2	1	3	6	8	4	9	5
1	6	9	5	2	3	7	8	4
2	4	8	6	9	7	5	3	1
3	7	5	8	1	4	9	2	6

163

3	6	1	9	5	7	8	4	2
5	7	9	4	8	2	6	1	3
4	2	8	3	1	6	9	7	5
7	9	3	5	2	4	1	8	6
1	4	6	7	3	8	5	2	9
8	5	2	1	6	9	7	3	4
6	1	7	2	9	3	4	5	8
2	8	5	6	4	1	3	9	7
9	3	4	8	7	5	2	6	1

164

1	5	8	4	9	7	2	6	3
3	2	4	6	8	1	5	9	7
7	9	6	5	3	2	4	1	8
5	3	7	9	6	4	1	8	2
9	6	1	8	2	5	7	3	4
4	8	2	7	1	3	9	5	6
6	4	5	3	7	9	8	2	1
2	7	3	1	5	8	6	4	9
8	1	9	2	4	6	3	7	5

165

7	8	5	4	9	1	6	3	2
1	9	2	7	3	6	4	8	5
4	6	3	8	2	5	9	7	1
9	5	6	3	8	7	2	1	4
2	4	8	6	1	9	3	5	7
3	1	7	5	4	2	8	9	6
8	3	1	2	7	4	5	6	9
5	2	9	1	6	8	7	4	3
6	7	4	9	5	3	1	2	8

166

5	6	3	2	1	4	9	7	8
2	8	9	7	5	3	6	4	1
7	4	1	9	8	6	3	2	5
9	7	4	6	2	5	1	8	3
6	2	8	4	3	1	7	5	9
1	3	5	8	7	9	2	6	4
8	5	7	1	9	2	4	3	6
3	1	6	5	4	7	8	9	2
4	9	2	3	6	8	5	1	7

167

1	3	2	9	7	4	5	6	8
5	6	9	8	3	2	7	1	4
8	7	4	6	1	5	3	9	2
7	9	6	3	8	1	2	4	5
4	2	1	5	9	7	8	3	6
3	5	8	2	4	6	9	7	1
6	1	5	7	2	9	4	8	3
9	4	3	1	5	8	6	2	7
2	8	7	4	6	3	1	5	9

168

8	6	2	1	7	9	4	5	3
9	4	3	2	5	8	6	1	7
7	5	1	4	3	6	2	8	9
3	2	4	5	8	1	9	7	6
1	8	6	9	4	7	5	3	2
5	7	9	3	6	2	8	4	1
4	1	7	6	2	5	3	9	8
2	9	5	8	1	3	7	6	4
6	3	8	7	9	4	1	2	5

169

2	3	7	9	5	1	4	8	6
4	8	5	3	6	7	1	9	2
1	6	9	8	4	2	3	5	7
5	4	6	1	2	9	8	7	3
8	7	1	4	3	5	6	2	9
9	2	3	7	8	6	5	1	4
3	9	8	2	1	4	7	6	5
6	1	2	5	7	3	9	4	8
7	5	4	6	9	8	2	3	1

170

1	8	4	5	2	6	3	7	9
9	5	6	3	8	7	1	2	4
2	7	3	1	9	4	8	6	5
4	1	8	9	7	3	2	5	6
5	9	7	2	6	1	4	3	8
6	3	2	4	5	8	9	1	7
8	4	5	6	3	2	7	9	1
7	2	9	8	1	5	6	4	3
3	6	1	7	4	9	5	8	2

171

4	3	5	2	8	9	6	1	7
6	7	1	3	5	4	8	2	9
8	2	9	1	7	6	3	5	4
3	1	2	5	9	7	4	6	8
5	9	8	6	4	2	1	7	3
7	6	4	8	1	3	2	9	5
1	8	3	7	2	5	9	4	6
9	5	6	4	3	1	7	8	2
2	4	7	9	6	8	5	3	1

172

3	9	7	5	1	6	4	8	2
6	4	1	9	8	2	3	7	5
8	2	5	4	3	7	9	6	1
7	8	4	2	5	1	6	3	9
2	6	9	8	7	3	5	1	4
5	1	3	6	4	9	7	2	8
9	5	6	7	2	8	1	4	3
1	7	8	3	9	4	2	5	6
4	3	2	1	6	5	8	9	7

173

3	8	6	5	4	7	1	2	9
9	2	7	8	1	3	5	6	4
4	1	5	9	2	6	8	3	7
1	6	2	4	7	5	3	9	8
8	7	3	6	9	2	4	5	1
5	4	9	3	8	1	2	7	6
7	3	8	2	6	4	9	1	5
2	9	1	7	5	8	6	4	3
6	5	4	1	3	9	7	8	2

174

2	3	4	6	9	7	1	8	5
7	8	5	3	2	1	6	4	9
6	1	9	8	5	4	7	2	3
5	7	2	9	6	3	4	1	8
3	6	1	2	4	8	5	9	7
9	4	8	1	7	5	3	6	2
1	5	3	4	8	2	9	7	6
4	2	6	7	3	9	8	5	1
8	9	7	5	1	6	2	3	4

175

2	7	1	5	9	3	4	8	6
5	9	4	1	8	6	7	2	3
3	6	8	7	2	4	5	9	1
7	4	5	6	3	8	9	1	2
9	8	3	4	1	2	6	7	5
6	1	2	9	5	7	3	4	8
8	2	7	3	6	9	1	5	4
4	5	6	2	7	1	8	3	9
1	3	9	8	4	5	2	6	7

176

3	4	6	9	1	7	5	8	2
8	9	2	5	6	3	7	4	1
7	1	5	8	4	2	9	3	6
6	5	3	7	8	4	1	2	9
9	2	1	3	5	6	8	7	4
4	8	7	2	9	1	3	6	5
5	3	4	1	2	8	6	9	7
2	7	9	6	3	5	4	1	8
1	6	8	4	7	9	2	5	3

177

6	3	8	9	1	4	2	7	5
9	2	7	6	3	5	1	8	4
5	4	1	2	7	8	3	9	6
3	6	9	5	8	2	7	4	1
1	5	4	7	6	3	8	2	9
8	7	2	1	4	9	5	6	3
7	8	3	4	9	1	6	5	2
4	1	5	8	2	6	9	3	7
2	9	6	3	5	7	4	1	8

178

1	5	8	6	2	4	7	9	3
3	9	6	8	1	7	4	5	2
2	4	7	5	9	3	6	1	8
8	2	9	3	6	1	5	7	4
7	6	5	2	4	9	8	3	1
4	3	1	7	8	5	2	6	9
9	8	3	4	5	6	1	2	7
5	7	4	1	3	2	9	8	6
6	1	2	9	7	8	3	4	5

179

7	8	6	5	2	1	3	4	9
1	3	4	7	8	9	2	6	5
9	2	5	4	6	3	7	8	1
3	4	2	6	9	8	5	1	7
8	7	1	2	5	4	9	3	6
6	5	9	1	3	7	8	2	4
2	1	7	8	4	5	6	9	3
4	9	8	3	7	6	1	5	2
5	6	3	9	1	2	4	7	8

180

6	9	1	4	3	7	2	5	8
7	2	8	6	5	1	9	3	4
3	5	4	8	2	9	6	1	7
4	8	7	1	6	2	5	9	3
1	3	5	7	9	8	4	2	6
2	6	9	3	4	5	8	7	1
5	1	3	9	8	4	7	6	2
9	4	6	2	7	3	1	8	5
8	7	2	5	1	6	3	4	9

181

8	6	4	5	2	1	9	7	3
7	2	9	6	3	8	4	5	1
3	5	1	4	9	7	2	8	6
9	8	2	3	4	6	5	1	7
4	1	6	7	8	5	3	9	2
5	3	7	2	1	9	8	6	4
6	9	3	8	7	4	1	2	5
1	4	5	9	6	2	7	3	8
2	7	8	1	5	3	6	4	9

182

4	3	2	7	5	6	8	9	1
6	5	1	4	8	9	2	3	7
7	9	8	1	2	3	4	6	5
1	7	4	5	3	2	6	8	9
2	8	9	6	4	7	5	1	3
3	6	5	8	9	1	7	4	2
9	2	6	3	7	8	1	5	4
5	1	3	2	6	4	9	7	8
8	4	7	9	1	5	3	2	6

183

3	4	5	7	2	6	8	9	1
7	8	2	9	1	4	6	3	5
1	9	6	3	5	8	2	4	7
5	6	1	2	3	7	4	8	9
9	2	8	6	4	5	1	7	3
4	7	3	1	8	9	5	6	2
2	5	9	4	6	3	7	1	8
6	1	7	8	9	2	3	5	4
8	3	4	5	7	1	9	2	6

184

3	1	2	9	8	6	7	4	5
9	8	4	1	5	7	3	6	2
6	5	7	2	3	4	1	8	9
2	7	9	6	1	3	8	5	4
8	6	5	4	7	2	9	1	3
1	4	3	5	9	8	2	7	6
5	9	6	8	2	1	4	3	7
7	2	8	3	4	5	6	9	1
4	3	1	7	6	9	5	2	8

185

4	1	8	2	9	3	7	6	5
3	2	7	8	5	6	4	9	1
6	9	5	1	7	4	2	8	3
8	4	3	6	1	7	9	5	2
1	7	9	3	2	5	6	4	8
2	5	6	9	4	8	3	1	7
7	6	1	4	8	2	5	3	9
9	3	2	5	6	1	8	7	4
5	8	4	7	3	9	1	2	6

186

5	2	8	4	9	1	7	3	6
1	6	9	2	3	7	5	8	4
3	4	7	5	8	6	9	2	1
9	3	6	7	1	5	8	4	2
8	1	4	3	2	9	6	7	5
2	7	5	8	6	4	1	9	3
6	5	2	9	7	3	4	1	8
4	9	3	1	5	8	2	6	7
7	8	1	6	4	2	3	5	9

187

1	4	6	2	9	7	5	3	8
3	9	5	1	6	8	4	2	7
7	2	8	4	5	3	1	9	6
5	7	9	8	4	6	2	1	3
6	1	4	3	2	5	8	7	9
2	8	3	9	7	1	6	5	4
9	6	2	7	1	4	3	8	5
4	3	1	5	8	9	7	6	2
8	5	7	6	3	2	9	4	1

188

2	6	9	5	4	1	8	7	3
7	8	3	2	6	9	1	4	5
1	4	5	8	7	3	2	6	9
8	7	4	1	3	6	5	9	2
3	2	6	7	9	5	4	8	1
9	5	1	4	2	8	6	3	7
6	3	8	9	1	2	7	5	4
4	9	2	6	5	7	3	1	8
5	1	7	3	8	4	9	2	6

189

4	2	3	7	8	1	6	5	9
6	1	8	4	9	5	3	2	7
5	7	9	3	2	6	8	4	1
8	4	7	5	3	2	1	9	6
9	3	1	8	6	4	5	7	2
2	5	6	1	7	9	4	3	8
3	8	5	2	1	7	9	6	4
1	9	2	6	4	3	7	8	5
7	6	4	9	5	8	2	1	3

190

6	4	9	3	7	8	5	2	1
2	8	3	9	1	5	4	7	6
5	1	7	2	6	4	9	3	8
8	3	1	7	9	6	2	5	4
9	2	5	8	4	1	3	6	7
4	7	6	5	3	2	8	1	9
7	5	2	6	8	9	1	4	3
3	9	4	1	2	7	6	8	5
1	6	8	4	5	3	7	9	2

191

7	8	1	9	2	6	4	3	5
2	5	6	3	1	4	9	7	8
9	4	3	5	7	8	1	2	6
8	7	4	1	3	9	5	6	2
3	9	2	6	5	7	8	4	1
6	1	5	8	4	2	3	9	7
5	6	9	7	8	3	2	1	4
4	3	8	2	6	1	7	5	9
1	2	7	4	9	5	6	8	3

192

1	2	5	8	7	3	9	6	4
3	9	7	1	6	4	2	5	8
4	8	6	2	5	9	7	1	3
6	1	2	4	8	5	3	7	9
5	4	3	9	1	7	6	8	2
8	7	9	3	2	6	1	4	5
9	6	4	7	3	8	5	2	1
2	5	8	6	9	1	4	3	7
7	3	1	5	4	2	8	9	6

193

2	6	1	9	5	8	7	4	3
8	5	4	6	3	7	9	2	1
9	7	3	2	4	1	5	6	8
7	8	5	3	2	9	4	1	6
4	1	9	5	8	6	2	3	7
3	2	6	1	7	4	8	9	5
6	9	7	4	1	5	3	8	2
5	4	2	8	6	3	1	7	9
1	3	8	7	9	2	6	5	4

194

7	1	9	5	6	2	3	8	4
5	2	6	8	4	3	7	1	9
8	4	3	9	1	7	5	6	2
2	6	5	3	7	1	9	4	8
3	7	4	6	9	8	2	5	1
1	9	8	2	5	4	6	7	3
9	3	1	7	8	6	4	2	5
6	8	2	4	3	5	1	9	7
4	5	7	1	2	9	8	3	6

195

4	9	6	1	5	3	2	7	8
7	2	8	4	9	6	3	1	5
3	5	1	2	7	8	9	6	4
6	4	7	9	3	2	8	5	1
1	3	2	6	8	5	7	4	9
5	8	9	7	1	4	6	3	2
2	1	5	3	6	9	4	8	7
8	6	4	5	2	7	1	9	3
9	7	3	8	4	1	5	2	6

196

1	9	8	6	4	7	2	3	5
6	3	5	9	1	2	4	7	8
2	7	4	8	5	3	1	6	9
8	1	7	2	9	6	5	4	3
4	2	9	3	7	5	6	8	1
3	5	6	1	8	4	7	9	2
5	6	1	7	3	9	8	2	4
7	8	3	4	2	1	9	5	6
9	4	2	5	6	8	3	1	7

197

2	8	6	4	1	7	3	5	9
3	5	7	2	6	9	8	4	1
4	9	1	5	8	3	6	7	2
7	2	4	3	5	1	9	8	6
8	6	5	7	9	2	4	1	3
1	3	9	8	4	6	5	2	7
5	1	2	6	3	8	7	9	4
9	4	3	1	7	5	2	6	8
6	7	8	9	2	4	1	3	5

198

2	9	3	6	4	8	5	1	7
4	6	1	9	5	7	2	8	3
5	7	8	2	1	3	9	4	6
6	5	4	3	2	9	1	7	8
8	1	9	7	6	5	3	2	4
3	2	7	4	8	1	6	9	5
1	8	2	5	3	4	7	6	9
7	3	6	8	9	2	4	5	1
9	4	5	1	7	6	8	3	2

199

9	7	8	4	3	2	5	1	6
2	6	3	9	1	5	7	4	8
5	4	1	7	6	8	9	3	2
7	5	2	3	8	9	4	6	1
3	8	4	1	7	6	2	9	5
6	1	9	5	2	4	8	7	3
1	9	7	8	5	3	6	2	4
4	2	5	6	9	1	3	8	7
8	3	6	2	4	7	1	5	9

200

1	9	8	2	5	7	3	4	6
7	4	3	8	1	6	9	2	5
5	6	2	3	9	4	1	8	7
2	7	4	6	8	3	5	1	9
8	5	6	1	2	9	7	3	4
9	3	1	4	7	5	2	6	8
6	2	9	5	3	8	4	7	1
4	1	7	9	6	2	8	5	3
3	8	5	7	4	1	6	9	2

THE TIMES

Su Doku

If you would like to receive email updates on the latest *Times* Su Doku and other puzzle books, please sign up for the HarperCollins email newsletter on **www.harpercollins.co.uk/newsletters**

Also available:

***The Times Su Doku* Book 1**
ISBN 0-00-720732-8

***The Times Su Doku* Book 2**
ISBN 0-00-721350-6

***The Times Su Doku* Book 3**
ISBN 0-00-721426-X

***The Times Su Doku* Book 4**
ISBN 0-00-722241-6

THE TIMES

Su Doku

The Times Su Doku **Book 5**
ISBN 0-00-722242-4

The Times Bumper Su Doku
(Books 1-3 in one volume)
ISBN 0-00-722584-9

The Times Killer Su Doku
ISBN 0-00-722363-3

The Times Su Doku for Beginners
ISBN 0-00-722598-9

The Times Su Doku
(mini format)
ISBN 0-00-722588-1

The Times Su Doku with Pencil
(mini format)
ISBN 0-00-722727-2